Democratising the Museum

Pille Runnel / Pille Pruulmann-Vengerfeldt (eds.)

Democratising the Museum

Reflections on Participatory Technologies

PL ACADEMIC RESEARCH

Bibliographic Information published by the Deutsche Nationalbibliothek
The Deutsche Nationalbibliothek lists this publication in the Deutsche Nationalbibliografie; detailed bibliographic data is available in the internet at http://dnb.d-nb.de.

Library of Congress Cataloging-in-Publication Data
Democratising the museum : reflections on participatory technologies / Pille Runnel / Pille Pruulmann-Vengerfeldt (eds.).
pages cm
Includes bibliographical references.
ISBN 978-3-631-64916-9 – ISBN 978-3-653-03972-6 (E-book) 1. Museums and community. 2. Museums–Technological innovations. 3. Museum techniques. 4. Museums–Educational aspects. 5. Eesti NSV Riiklik Etnograafiamuuseum. I. Runnel, Pille. II. Pruulmann-Vengerfeldt, Pille.
AM7.D455 2014
069--dc23

2014005518

This book was published with the support of Estonian Science Foundation grant no. 8006 and the Estonian National Museum.

English language editor: Daniel Edward Allen
Copy-editors: Tuuli Kaalep, Ivi Tammaru
Layout: Tuuli Kaalep, Ivi Tammaru
Cover photo: Marko Mäetamm, „Colourful friends“ (1995)

ISBN 978-3-631-64916-9 (Print)
E-ISBN 978-3-653-03972-6 (E-Book)
DOI 10.3726/ 978-3-653-03972-6

PL Academic Research is an Imprint of Peter Lang GmbH.

Peter Lang – Frankfurt am Main · Bern · Bruxelles · New York · Oxford · Warszawa · Wien

www.peterlang.com

Table of Contents

IV Digital Technologies for Participation

V Conclusion

Authors

Agnes Aljas is a research secretary at the Estonian National Museum and a PhD student at the Institute of Social Studies, University of Tartu. Her PhD research focuses on changes in the concept of culture. Currently, she is participating as a researcher in the following research projects: “Changing Cultural Dispositions of Estonians through Four Decades: from the 1970s to the Present Time”, “Developing Museum Communication in the 21st Century Information Environment”, and “Actual Complexity of Cultural Communication and Methodological Challenges of Cultural Research”.

Nico Carpentier (PhD) is Associate Professor at the Communication Studies Department at Vrije Universiteit Brussel (VUB – Free University of Brussels) and Lecturer at Charles University in Prague. His theoretical focus is on discourse theory; his research interests are situated in the relationship between the media, journalism, politics and culture, focussing on social domains such as war and conflict, ideology, participation and democracy.

Krista Lepik (PhD) is a lecturer at the Institute of Social Studies, Faculty of Social Sciences and Education, University of Tartu. Her research interests cover governmentality and cultural participation in museums and libraries, and information literacy education in an academic environment.

Linda Lotina is a PhD student at the Institute of Social Studies, Faculty of Social Sciences and Education, University of Tartu, and a lecturer at the Vidzeme University of Applied Sciences. Her research focuses on participation issues in (Latvian) museums.

Pille Pruulmann-Vengerfeldt (PhD) is an associate professor at the University of Tartu, Institute of Social Studies, and a researcher at the Estonian National Museum. Her interests are Internet user typologies, user-friendly online spaces as possible venues for participation, and participatory applications for organisations. She is leading and participating in several national and international projects.

Pille Runnel (PhD) is a research director of the Estonian National Museum. She has worked as a researcher at the Institute of Journalism and Communication, University of Tartu and is carrying out joint research projects with the Institute in the areas of generations in the emerging information society, and museum communication in the context of the information society. Her research interests involve ICT and cross-media usage, and audiences (governance, participation, the digital divide). She has been part of the EU Kids Online research initiative, as well as several research projects on transforming media usage. Her research also involves users and digital cultural heritage, visual anthropology and new museology. She is responsible for the research agenda and the production of future permanent exhibitions at the Estonian National Museum.

Taavi Tatsi (PhD) is an independent researcher formerly affiliated with the Estonian National Museum. His PhD research focused on the transformations of museum-embedded cultural expertise, specifically how audience participation affects museum curatorship and collections. His research is part of the "Developing Museum Communication in the 21st Century Information Environment" research project.

The Challenge of Democratising the Museum

Pille Pruulmann-Vengerfeldt

Pille Runnel

1. The project

Democratising the museum is a collection of studies looking at how participation can support museums in the process of becoming more open. We look at the open museum as sharing power with its visitors and stakeholders, but also negotiating professionalism and the role of the museum in a modern society. The societal challenges that inspire this book are on the one hand related to the need to reinvent the notion of democracy in today's political crisis, to open up the concept and bring it out of the strictly institutional level of politics. On the other hand, the challenges are to do with increasing technologisation of society, ways of sharing information, communicating and networking with the public. While often the technologies are seen as the solution to the problem of democratisation, the key thesis of this book is that the way technologies are used are far more important. Hence the concept of participatory technologies is under investigation.

This is a project book. It draws upon a five-year research project called "Museum Communication in the 21st Century Information Environment", carried out at the Estonian National Museum. At the same time, it is much more than a book about one museum research project, one team and one set of participatory initiatives. The articles collected here reflect journeys of hope and expectation about museum development, audience engagement and the role of technology in these processes. This book looks at successful ways in which museums use different techniques and technologies to foster museum communication, especially audience participation. But we also talk about some of our failures.

Curiously, the project is not set in traditional museology. Our research started by identifying a research issue and looking at a museum in dire need of research to help it face multiple challenges. We also started by forming a team of researchers who were both eager and interested to bring together various disciplines to help the Estonian National Museum in the transformations it was going to face, and to find new ways to analyse how museums operate and interact with their audiences. The project partners and authors of this book come from a variety of disciplinary backgrounds within media and new media studies, media sociology, ethnology and museology, communication studies, information and technology studies and finally, democracy and participation studies. Moreover, many of the partners have been affiliated both with the museum and with the university at the same time and were simultaneously studying the museum and its processes as well as trying to shift some of the key understandings of how Estonian National Museum as an institution should operate.

What kind of promises does this interdisciplinary approach make to the readers of this book? As a number of these articles or their earlier versions have been previously published elsewhere, we can use this introduction to bring together the conceptual contributions we believe this interdisciplinary background provides. Interdisciplinarity combined with strong theoretical background can be very beneficial for transdisciplinary developments (Pruulmann-Vengerfeldt, 2013). We use the theoretical concepts and/or the methods from different disciplines and can thus shed new light and new understanding on an area in which study is not new. This means that some of the statements present here might be obvious and familiar to some disciplines while we believe that they have new and important contributions to make in other fields.

This book brings together a set of case studies conducted mainly in one particular museum, contextualised with some international case studies from the members of the same research team. Being based on a thorough theoretical and analytical framework, this book could be a useful handbook to anyone interested in involving audiences when making any kind of cultural institutions more participatory or more audience centred. We hope that strong analytical roots in the workings of one particular museum means that the discussions are strongly connected to the real museum context and that this book can thus also serve as an asset for other memory institutions, such as galleries, libraries and archives also struggling with the same processes. As this book combines research with practice and action research experiments conducted throughout the research period, readers with professional backgrounds might find that the articles collected here inspire and offer opportunities to learn from the challenges we encountered throughout our journey. At the same time, the very same point of departure of combining theoretical ideas from different research disciplines and trying them out in the museum setting, makes this collection of articles a valuable resource for researchers and students working in the areas of museology and museum audiences, and in the fields of communication and audience research more generally.

2. The journey

When we embarked upon the journey of proposing a research project more than five years ago in 2008, we had some hopes and dreams. When we started looking at museums as sites for communication and participation and ways in which technology can foster these activities, we were not quite sure what to expect. The original "Museum Communication in the 21st Century Information Environment" research project was founded on the expectation that construction of the new Estonian National Museum would soon be finished, and that the overall process of research would emerge along with finalisation of the first museum building for the hundred-year-old Estonian National Museum. When we started outlining the research plan, the building process was at the stage of summing up negotiations between architects and the engineers regarding the building project, whereas the museum had hardly worked with the detailed floor plans according to the latest understandings of how a contemporary museum would operate in a space specifically designed for a museum. The building along with its contents was expected to be finished in 2012. We hoped that – with the unique opportunity of seeing, analysing, recording and participating in the process of constructing a new building for a museum that had already existed for over 100 years – we would be able to bring academic knowledge to the process of making and reinventing a museum. We embarked upon the project with an enthusiastic hope that by evolving emerging new technologies, we might be able to open the museum and the building process to the wider public, involve people in the debates about future museums and discussions over the ways a renewed museum would operate in terms of collections, content, exhibitions and programmes.

It could be said that the new building would have brought a new museum as well as new museology. However, the funding of the building was declined unexpectedly around the hundredth anniversary of the museum, followed by several months and eventually years of uncertainty and negotiations of securing other source of funding in order to continue building the museum. In 2013, the final year of our research project, there is no new museum building yet, although the corner stone was finally set in the spring of 2013. The new building of the Estonian National Museum is currently on its way and should open its doors in 2016. Hence it can be said that the new house that we as a research team hoped to fill with the discussions and debates emerging from the participatory framework, is still largely on the drawing boards. At the same time, we have had a unique possibility to send the museum on its way to new physical settings and on its way to becoming a renewed organisation as experimental laboratory for participatory initiatives. We were also able to use our double vision as academics and museum professionals to learn from these initiatives and study them from the academic viewpoint.

This book reflects our academic journey. It is a way to summarise some of our findings as well as some of our enthusiasm and optimism. The delay in the new museum building changed the opportunities open to the research project, influencing the project to evolve in different ways. Eventually, it became a study of museum communication in three different areas, looking at the museum institution from the aspect of public debate, from the aspect of museum content and from the aspect of the museum-making process. We aimed to understand the current status of things and at the same time design interventions that would change the way in which a particular museum was perceived. In this process, the focus shifted away from understanding the current status, as many of the project members were working at the museum and felt that the current status was less relevant. We set out to change not only the public perception of the museum but also the perception from inside the museum. Applying participatory action research meant that many of the research project members were at the same time studying the museum and its processes as well as trying to shift some of the key understandings of how museum should operate.

The project has been unique in its magnitude. Over the five years, by involving key personnel from the museum, research interventions were carried out, involving different aspects of museum work in order to analyse and change the ways in which the museum operated at different levels. We became involved in exhibition making, collecting and communication with communities.

In the initial stages of the project we were dreamers. We wanted to make big changes in how the museum perceived its audiences and in how audiences perceived the museum. However, the project soon encountered resistance, misunderstanding and miscommunication inside the museum as well as from audiences, and at the end of the day we learned a lot from these experiences. Some of these hardships are reflected in this book. In having tried the ideas out in a real museum setting with very limited resources we believe that this book could be a valuable resource to anyone interested in making museums more participatory and more audience centred.

The double involvement, with many team members being involved with both university and museum work, created a unique sense of vision with the team having to take the perspective of the researcher and participant at the same time. To engage more perspectives, the project team slowly grew, which also influenced the outline of the book. The four sections of the book look at the theoretical foundations of participation within the museum, the methodological how-tos for conducing and analysing a participatory project, analyses of the roles of visitors/audiences and cultural professionals, and finally the application of digital technologies in museum communication processes. The book not only focuses on the Estonian National Museum; five of the articles in this collection deal with the questions of participation technology, cultural professionalism, audiences and the role of the museum from a wider perspective.

3. The concepts

We use the different chapters to give meaning to the notion of participation. While earlier studies on museum participation are more strongly influenced by somewhat idealistic and normative expectations that participation is both good and necessary for the museum, later analyses have given the notion a slightly more critical stance, although still supporting participation as an important way for audiences to forge relationships with the museum while perhaps being a bit more realistic about the massive investment good audience participation requires from all parties.

Central to this book is the conception of audiences. Here we borrow the conceptual framework from media studies, rather than museology. Using 'audience' as a primary notion rather than the more familiar notion of 'visitors' throughout the articles is a conscious choice, as we take a stance in the active audiences paradigm where people receiving messages are seen as an important part of production of meaning. 'Visitors' come and go and leave a mark only when invited to write something in the guest book. However, viewing people communicating with the museum as audiences for a variety of museum actions, we join the tradition of media studies, in which audiences are actively part of meaning making (Schrøder, 2009) and in the case of participatory activities, audiences become produsers (Bruns, 2006). Still, when applying the notion of 'audiences' as a kind of umbrella term, many other labels are used in this volume. We speak of audiences, visitors, users, participants, communities, while the underlying principle remains the same – all are understood as active and engaged roles. These people can choose to opt in or opt out of the discussions taking place at the museum. Therefore, even if one attends a museum as a passive visitor just viewing one or other exhibition, the role of viewing and never participating is still an actively (although not always consciously) chosen position.

As this collection of articles deals with the field of museum communication, it also inevitably looks at museum workers as cultural professionals. A number of articles in this volume analyse their professional struggles, brought on by challenging changes in society, technology and the role of museums. Here the parallels of media production and thus the relationship with the theoretical framework of media studies are much more marginal.

At the beginning of the research project, the participatory technologies in the title of this volume were mainly seen as internet related technologies. However, in the course of this research we learned that technology is only part of the equation. Participatory technologies used in the course of this project range from pens and papers to internet, mobile phones and social networking technologies. As the example of handicraft hobbyists (see same volume: Lepik, Pruulmann-Vengerfeldt, 77–88) demonstrates, knitting needles, crochet hooks, carpentry and smithing can also be used as means of fostering participation and engagement. Participatory technologies go beyond technical devices and should rather be understood as an innumerable range of approaches, activities and views that museums can employ to foster museum engagement among their audiences.

Our conceptualisation of the museum within the boundaries of this research can be considered rather old-fashioned and classical than innovative and boundary blurring. While we are aware of many new initiatives, pop-up museums, newseums and other experiments that change the way in which audiences understand and experience the museum institution, our take on analysing museums for this particular volume is more conservative. In this book, discussions of the changing museum institution stem from investigation of the institutionalised, nationally funded museum and the changes that have been forced upon it in the face of societal shifts. Thanks to this institutional focus we see the analysis to be still relevant and applicable to other contexts, such as galleries, libraries or archives which are also struggling with the same processes: trying to remain relevant without completely giving up their institutional conceptualisation.

Museums along with other cultural institutions struggle in the intersection of political, cultural and economic fields both when it comes to the definition of professionalism, the division of resources and competition for audience attention (the notion of fields is inspired by Bourdieu (1998), see same volume: Pruulmann-Vengerfeldt, Runnel, 35–53 for extended review of fields in the museum context). These fields have different operational logic that implies different demands for audience engagement. The articles in this book look less at the commercialisation, market-drivenness and playfulness of the museum experience, which are currently drawing wide attention. At the same time especially in the section of the book where we look at the professional challenges of museum workers, these tensions between the operational logics of the cultural, economic and political fields are prominently noticeable in the ways in which they influence willingness to test out participatory technologies. Hence, it is also important to acknowledge the political and economic fields when focusing primarily on a museum as a field of cultural communication.

4. The new era

With this book, we aim to make an important contribution towards the new era of museum studies. Today, in a situation where museums are competing for the most limited resource of all – the attention of the people – museum studies need to look carefully at what is done in other disciplines to understand better the dynamics of museum communication. Thus, the contribution this book hopes to make, is strongly rooted in the interdisciplinary background of the research team who bring together conceptual contributions from a variety of fields. Some of the aspects of the on-going change are discussed in contributions to the book *The Digital Turn: User's Practices and Cultural Transformations* (Runnel et al., 2013), particularly regarding the importance of the digital technologies, foregrounding not necessarily the technologies themselves, but the people as part of the digital turn (Pruulmann-Vengerfeldt et al., 2013). While the title of this book: *Democratising the Museum: Reflections on Participatory Technologies*, seemingly emphasises the technical component of the democratisation process, it shows that technologies mean more the ways of doing this and less the actual nuts and bolts or screens and hard drives of applied digital technologies (pen-and-paper initiatives that support democracy building by giving a voice to the people are also viewed as part of democracy-building).

More generally, the change we are talking about is a turn towards a communicative museum where the new technologies introduced are first and foremost communication technologies, enabling dialogue, interaction and power-sharing. One of the authors in this book has elsewhere highlighted the shift in contemporary museum communication from the monovocal, in which the museum speaks in a single voice to the masses, to multivocal, in which the museum makes space for other speakers (Tatsi, 2013). In this understanding, the museum becomes space not just for dialogue between the museum and its audiences, but rather a space for discussions and interactions. In order to understand this aspect and role of the museum more thoroughly, we also need to incorporate communication research in the uniquely rich mix of museum studies.

With the communication dimension becoming increasingly important in studying the museum field, museum studies 2.0 is also embarking on the notion of a social museum. Participatory engagement that is more than just contributing to the museum when asked, recognising the importance of social interaction not only with the museum and its contents, but also with museum-goers amongst themselves. This does not mean that the museum should be replaced with a bazaar and that everyone should be left guessing as to whose voice can be heard today. Rather, this kind of transformation to a social museum simply refers to the widened repertoires of the museum, understanding of which also calls for a change in museum research.

Already in 1989, when Vergo called for new museology, the role and purpose of the social dimension within the museum was under consideration. This collection contributes to the discussions of these considerations, analysing the framework for such considerations in the first section, changes in understanding of the people in the museum in the second section, analysing the shifts in professionalism in the third section and looking at the technology in the new museum in the final section. In a way, this is also the order in which these challenging aspects come about for the museum. The larger societal considerations and the role of the people for museums are more relevant than the accompanying technological changes, which do play a role in enabling some of the changes, but by no means define them. We hope that this book will be a valuable resource that will support development and analysis of the museum in these changing contexts. Especially as the museum is an ever-changing research object, we see new museology as still needing to be defined, conceptualised and studied, even more than twenty years after the statement made by Vergo in *The New Museology* (1989).

Acknowledgements

The publishing of this article was supported by the Estonian Science Foundation grant no 8006 and the Estonian National Museum.

References

Bourdieu, Pierre (1998). *Practical Reason: On the Theory of Action*. Cambridge: Polity Press.

Bruns, Axel (2006). Towards Produsage: Futures for User-led Content Production. In Faye Sudweeks, Herbert Hrachovec, Charles Ess (eds.). *Cultural Attitudes towards Technology and Communication 2006*. Murdoch: Murdoch University, pp. 275–284.

Pruulmann-Vengerfeldt, Pille (2013). King of the Hill? Seeking the New in Audience Research. *The Communication Review*, 16 (1–2): 92–102.

Pruulmann-Vengerfeldt, Pille; Pille Runnel, Marin Laak, Piret Viires (2013). The Challenge of the Digital Turn. In Pille Runnel, Pille Pruulmann-Vengerfeldt, Piret Viires, Marin Laak (eds.). *The Digital Turn. User's Practices and Cultural Transformations*. Frankfurt am Main: Peter Lang Verlag, pp. 7–12.

Runnel, Pille; Pille Pruulmann-Vengerfeldt, Piret Viires, Marin Laak (eds.) (2013). *The Digital Turn. User's Practices and Cultural Transformations*. Frankfurt am Main: Peter Lang Verlag.

Schrøder, Kim (2009). Audience Theories. In Stephen W. Littlejohn, Karen A. Foss (eds.). *Encyclopedia of Communication Theory*. Los Angeles: Sage, pp. 63–68.

Tatsi, Taavi (2013). *Transformations of Museum-embedded Cultural Expertise*. Dissertationes de Mediis et Communicationibus Universitatis Tartuensis, 18. Tartu: University of Tartu Press.

Vergo, Peter (ed.) (1989). *The New Museology*. London: Reaktion Books.

THEORISING AND ANALYSING PARTICIPATION

Who Authors the Nation? The Debate Surround the Building of the New Estonian National Museum

Pille Runnel

Taavi Tatsi

Pille Pruulmann-Vengerfeldt

Estonia, the northernmost of the three Baltic states, regained its independence as nation state in 1991, which it had lost with the Nazi-Soviet pact in 1939 and the Soviet occupation of Eastern and Central Europe during the Cold War. Its earlier period of independence, from 1918 to 1940, was short lived and from the outbreak of hostilities in the Second World War, the country suffered the loss of a significant portion of its ethnic population through emigration and deportation, only to have it replaced by Soviet migrants and military units. For the majority of ethnic Estonians much of the twentieth century was an immensely traumatic experience and with the restoration of independence in 1991 came the hope that the nation might pick up from where it left off more than fifty years before. This chapter concerns the Estonian National Museum which was created as a part of the national movement in 1909 and which then established itself as an important symbol of national memory and identity. In the early 1990s, in the 'period of national awakening' when the country underwent major reform, there developed the idea of building a new Estonian National Museum. It arose in that period of hope and ideals, which straddled the moment when independence returned but it soon found itself locked in a period of pragmatism and economic reality (Runnel et al., 2009). Indeed, with large-scale economic turmoil sweeping Europe in 2009, doubts and questions began to emerge concerning the future of the project to build the museum and the value of a national museum to modern Estonian society.

The debate surrounding the erection of the new Estonian National Museum took place within what Michael Kennedy (2002) called the 'transition culture' that enveloped post-communist nations in Central and Eastern Europe in the late 1980s and early 1990s. The initial liberation was a large-scale grass-root civic initiative, which soon was channelled into forming of the institutional structure of the new republic. It involved fundamental changes and efforts to build social order, technology and infrastructure and included disappointment and hard times when the economic recession hit the new republic. Lauristin and Vihalemm (2009) interpret this phenomenon as a field of mediation, where external demands from the powerful international agencies are, through a specific 'learning process', turned into the value preferences and codes of behaviour of the actors within the transforming state; external definitions of 'success' and 'failure' are by these means 'naturalised' in evolving values and practices. The resulting blurring of internal and external agendas in this field of mediation has certainly been apparent in the processes involved in the establishment of new cultural institutions (Tali, Pierantoni, 2008a). Recent Western interest in major museum building projects, for example, has resulted in much debate about the spatial conceptualisation and impact of these institutions (Stead, 2004) which has spilled over into Estonian discussions of its new national museum. These have fostered expectations which variously see the museum as part of the rapidly developing creative industries, a contributor to the knowledge infrastructure, and a home for civic and ethnic nationalism.

Estonian conceptualisation of the museum and its social purpose was also undergone change. On the one hand, there has been the necessity to define identities at a time of rapid change by locating and securing old values and repaying history's debts. The latter, in particular, had wide support, as the social disruption caused by the Soviet occupation destroyed lives, artefacts and institutions; now public opinion was in favour of re-establishing those things lost in both the cultural and political arenas. The reinvention of the museum has been closely connected to these questions of collective memory and collective identity, which have in turn also been affected by a return to the European fold.

This study has emerged from the fields of media and communication research and the anthropology of cultural production. It has applied an ethnographic methodology involving participant observation in meetings and in regular working practices, and analysis of different media sources. It draws upon Peterson's (2003: 177) Bourdieuian analysis of production culture: "a complex network of relations between various institutions and agencies that have various kinds and degrees of power over aspects of media production", which involves "an ongoing construction of social actors working in it." Of particular interest to us has been the role of authorship and its ownership amongst various actors, and the manner in which audiences are permitted to participate in decision making and to which extent this participation is only token or pseudo-participation (Carpentier, 2007). Participation here can arise from an actor's membership of various communities.[1] Thus, we were also interested in how audience engagement is envisaged within the field of production, the composition of that audience and how it participates in constructing the museum's narratives. We had in mind van Mensch (2005) assertion

1 Graffman, Katarina (2004). *The Cruel Masses: How Producers at a Swedish Commercial Television Production Company Construct Their Viewers.* Available at: http://www.media-anthropology.net/graffman_thecruelmasses.pdf (accessed May 15, 2008).

that museums need to be laboratories and meeting points for discussion and new initiatives. In Estonia recently released from the shackles of the Soviet system this was for many a very novel conceptualisation of the museum.

1. Authoring the nation

The location of the original museum building became historically significant to Estonians for many reasons. Raadi Manor (Figure 1), on the outskirts of Tartu, the second largest town in Estonia, prospered under the Baltic-Germans in the nineteenth century and enabled the owner to develop his interest in flying. This resulted in the establishment of a small airfield, which decades later made the area militarily attractive, and which would place an important role in the future conceptualisation of the museum. During the nationalist reforms of the early years of independence in the early twentieth century, the manor was given over to the University of Tartu, which agreed to share part of it with the Estonian National Museum the manor was given over to the Estonian National Museum, which was actively seeking a building of its own at that time. The museum situated itself in the less than ideal main building, where a popular permanent ethnographic exhibition focusing on peasant culture opened in the early 1920s. The manor's extensive grounds became a favourite spot for excursions and walks, and with its museum established Raadi as an important symbolic place for the whole of Estonia. How-

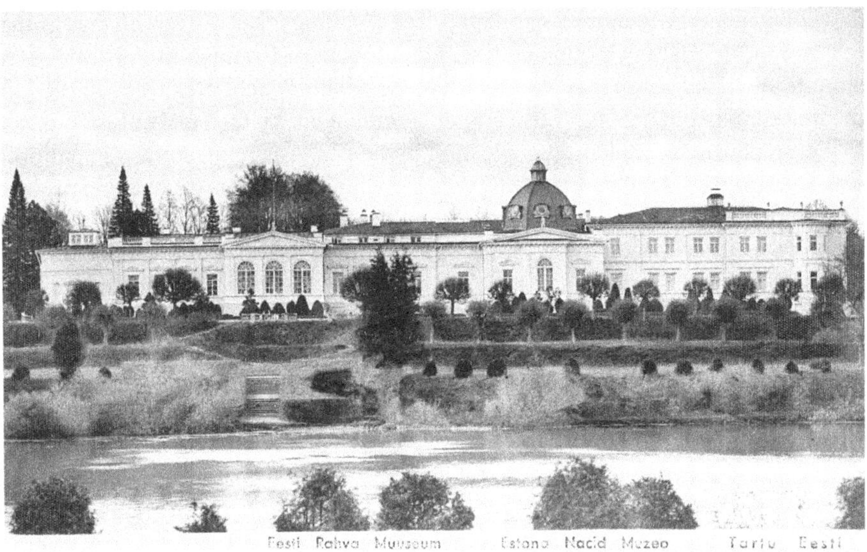

Figure 1: Raadi Manor, 1938. Photo: Eduard Selleke, Estonian National Museum

ever, the outbreak of the Second World War saw the evacuation of the collections and Soviet bombing of Nazi troops quartered there, left the building in ruins and the whole area in the hands of the Soviet air force.

For Estonians, this military and destructive influence on the Raadi area, which turned Tartu into a closed city, was a poignant metaphor for the overall situation Estonia had found itself in during the fifty years of Soviet rule. In April 1988, at the beginning of the public movement against the Soviet occupation, a demonstration was held calling for the Estonian National Museum to be returned to Raadi. The following year, the Soviet army made a partial withdrawal from the property and land was appropriated for the 'Estonian National Museum at Raadi'.

From 1993 to 2005, discussion of the future location of a more permanent museum building revealed no consensus even within the museum itself. For many Raadi was in all senses a ruin: polluted both physically and symbolically.

In 1994, early in the transition period, an exhibition on Estonian culture was established in a temporary building at the centre of the town. The exhibition, in keeping with the dominant sentiment of the day, was deliberately ethno-romantic with its focus in the display of nineteenth-century peasant life. The curator responsible for the exhibition reflected a decade later that the role of the Estonian National Museum was to maintain Estonian identity, and its task was to find different ways to accomplish this objective.[2]

The Raadi area was not entirely forgotten though and a competition for future development of the manor was arranged. The resulting ideas were as diverse as the museum of the Baltic German culture, the museum of Estonian life histories and a multifunctional cultural centre. In 1998, the central manor area was returned to the museum by order of the Tartu city government but by then there existed plans to erect a new museum building in centre of the city. An architectural competition was held for the Estonian architects, which resulted in choosing a winning project and preparing the building site. However, delays in the construction work and a pressing need for space,[3] led to the museum building its stores at Raadi, beginning the work in 2000. Building the storages at the Raadi certainly facilitated turning the opinion in favour of building the new museum at Raadi slightly later, following a number of newspaper articles published between 2001 and 2004, which reawakened a nostalgic reflection on that special place in the minds of an older generation of Estonians: "From now on, the museum's Raadi-era started, the re-

2 This point was made in round table discussions in 2006 about the different exhibitions in Estonian museums.

3 The museum needed to remove its collections from temporary storages in different churches at Tartu and return the buildings to the congregations, as the buildings were leased/rented until 2005.

membering of which brings a sparkle in the eye of the older generation".[4] In the national newspapers, the new museum became a symbolic object of national importance, and the Estonian government's decision at 2003 to rebuild the museum in the Raadi area[5] was interpreted as a triumph of justice. The museum now became the material, bodily manifestation of the nation:

> The return of the Estonian National Museum to Raadi is also the return to the home of the nation, which should give us certainty of our national cultural survival. There is no future without the past for a nation, let us take care of the past.[6]

In 2005, an international architecture competition was announced.[7] Estonian historian, Marek Tamm (2005),[8] described the Estonian National Museum as 'one of the most important displays of Estonian culture', but emphasized in the introductory text to the competition that the museum's identity was bound to change:

> It is clear that today the golden age of nationalism is now behind us, and that a museum dedicated to displaying objects from a single nationality (and other Finno-Ugric nations) does not fit in with the terrain of other European museums. The ENM's future should be, before all else, to operate as an ethnological museum with an open spirit and a diverse collection, which will gather, store, research, and display different cultural inheritances, not so much on a national as a scientific basis.[9] But also, the ENM should preserve the knowledge of its historical role in the creation of the Estonian nation and through its collection investigate and display this role.

4 Mikelsaar, Raik-Hiio (2003). Rajame Raadile estoloogiakeskuse ja mini-Eesti? [Developing estology centre and miniature-Estonia to Raadi?] *Tartu Postimees* 98 (1438), 23.5.2003, p. 2.

5 After the thorough discussions in the councils for museums, architecture and heritage protection, the councils decided that the Raadi area is the most promising one in terms of the possible future developments of the museum and they decided to support the erection of the new museum building over there. At July 31st, 2003, the ministry of culture confirmed the decision.

6 Ja saagu muuseum! [And let there be a museum!] *Postimees*. Available at: http://vana.www.postimees.ee/index.html?op=lugu&id=110181&number=870&rubriik=6 (accessed March 20, 2010).

7 The international, public architectural competition started at June 22th, 2005, organised jointly by the ministry of culture and the Estonian Architects Union.

8 Tamm, M. (2005). *The Status of the ENM within Estonian Culture.* http://www.museum-competition.org/en/museum (accessed October 27, 2008).

9 Native English speakers should note that in continental Europe the term 'scientific' is not used in the same narrow and particular sense in which it is applied in the UK, USA and elsewhere. Here it refers to rigorous academic study and thus embraces such subjects as history without implying old-fashioned historiography or a narrowly 'scientific' conceptualisation.

The competition attracted Europe-wide interest and, quite unexpectedly, settled on a winning concept which positioned the museum building as extension to the runway on the airfield at Raadi. The concept, developed by three young architects from Paris, Dan Dorell, Lina Ghotmeth and Tsuyoshi Tane, titled "Memory Field",[10] completely ignored the common public understanding of the museum as the repository of the country's romantic peasant past. Instead, the winning design attached itself to wider contemporary historical debate and something the media had not even considered when discussing the general vision for the museum. It had seemed unthinkable that the Soviet occupation could be part of the discourse on Estonian identity. The judging committee remarked:

> The ideological premise behind this entry is somewhat unexpected and surprising given Estonia's dramatic recent history – the devastating Soviet occupation lasting more than half a century. This history cannot and must not be banished from the nation's memory by denying the traces still present; rather, these traces should be given a new meaning that inspires hope. This is a design that opens up discussions. (ENM 2006)

Dan Dorell, one of the architects of the design, said that their inspiration had come from Berlin where the Berlin Wall had been transformed from being a metaphor for Eastern Block repression into art that integrated history into the urban space giving it new meanings.[11] Contemporary Berlin now had numerous objects and spaces that connected the city and nation to its recent history in the lived environment.

Andres Kurg, an Estonian architectural historian, saw in the new museum's design a tension between 'appropriate' and 'inappropriate' elements, as they might be termed in the collective memory of Estonians. He noted that for these Western European architects, the airfield signifies the polarity of the Cold War, rather than the occupation of Estonia: "For locals it also signifies pollution, confinement, Dzhokhar Dudayev[12] or something else."[13] Kurg saw the airfield (Figure 2) as a scar left from a century of modernisation which was situated in the midst of

---

10 The architects, originally from Italy, Lebanon and Japan, have later on established themselves as an architectural bureau „Dorell.Ghotmeth.Tane“ with an office in Paris.

11 Aesma, Madis (2006). Võitjaarhitekt Dan Dorell: Raadist võib saada põnev paik. [Winner-architect Dan Dorell: Raadi could be exiting place.] *Postimees*. Available at: http://tartu.postimees.ee/170106/tartu_postimees/uudised/188990.php (accessed February 27, 2008).

12 Dzhokhar Dudayev (1944–1996) was the first president of Chechnya and before that a Major-General of Raadi air base in the last years of the Soviet Union. His actions during that period are seen to have favoured the Estonian nationalist movement over the Soviet authorities.

13 Kurg, Andres (2006). Hoolikalt polsterdatud arhitektuurivaidlus. [Carefully padded architectural debate.] *Eesti Ekspress*. Available at: http://www.ekspress.ee/viewdoc/EC4C730C771D9B82C22570FF00619456 (accessed October 27, 2008).

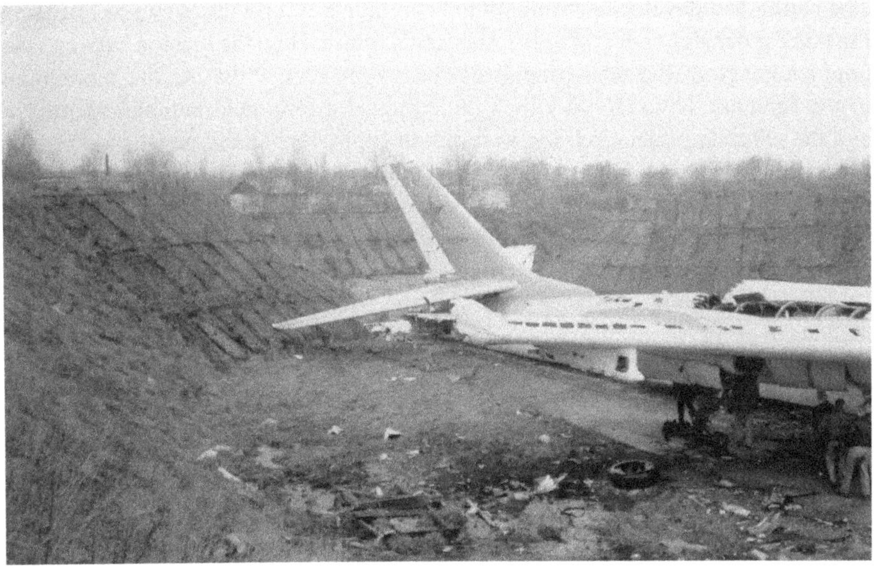

Figure 2: Raadi Airfield. Photo: Museum of the Estonian National Defence College

conflicts and their after effects but yet which were set to continue to shape the modern nation. He felt that collecting this history together and explaining it was very much a task for the future Estonian National Museum.

The architects understood that recent historical events evoked strong emotions in contemporary Estonians but their desire was to locate a physical and mental space which opened up modes of meaning-making beyond that of explicit condemnation and the eradication of all of residues of the Soviet past. It was seen as an important new role for the Estonian National Museum. The museum also had to recognise that it had entered an increasingly heterogeneous world, that it existed in a state now fully engaged in Europe and with global socio-cultural and intellectual trends. This, however, brought its own challenges, for Estonians also needed to restore and consolidate their own ethnically-based sense of nationhood; the world was open to them after 50 years of closure, but they still had work to do at home.

Not everyone saw the plan for the museum positively. Some felt it perpetuated the occupation and that the imposition of these connections on the very fabric of the national museum only served to humiliate and undermine Estonian identity.[14]

14 Hallas-Murula, Karin (2006). Võidutöö tekistas masendust. [Winning work depressed.] *Postimees*. Available at: http://tartu.postimees.ee/180106/tartu_postimees/arvamus/188991.php (accessed October 27, 2008).

The nation had developed strong online communities, and the Web soon became the hub for debates concerning the museum's future. Here the tension between the popular image of the ethnographically-based museum – that secure monument to the romantic national era which underpinned a long-held national identity – and the winning project, relying as it did on controversial and negative symbols, produced considerable argument. This popular debate indicated that, while the project had its supports, the critical voices in the public opinion became most well heard. One commentator wrote, outraged at the architects' suggestion that the project offered therapy for Estonians, 'this Soviet airfield runway is symbolic suicide... Frustrating ignorance regarding the common cultural memory!!'[15] The idea seemed an imposition which ignored the public's collective memory altogether. Several groups felt that the museum's representations of the past should be a string of 'beautiful events and secure symbols', which ignored bloodier history. The airfield plan served to glorify the occupation and open up old wounds in a kind of 'psychoanalysis' of the past. Others thought such attitudes were akin to burying one's head in the sand. The museum had already embarked capturing history in all its forms and circumstances:

> Estonian cultural heritage, for the documentation and preservation of which the Estonian National Museum has been established, as far as I know, is a way more diverse phenomenon than Jakob Hurt[16] and the Finno-Ugric. The museum is for example, actively collecting material from the Soviet period… If the ENM building adopted the calming and comforting mode of the barn swallow,[17] it would mean that we would still be able to identify ourselves only with the secure self-image and the image of history. Which is unfortunately just an illusion.[18]

---

15 Hallas-Murula, Karin (2006). Võidutöö tekistas masendust. *Postimees*. Available at: http://tartu.postimees.ee/180106/tartu_postimees/arvamus/188991.php (accessed October 27, 2008). Reader's comments on the article.

16 Jakob Hurt (1839–1907), one of the central figures in the Estonian national awakening movement, initiated a massive folklore collecting campaign in 1888, thus paving the way for the collecting activities that led to the founding of the Estonian National Museum in 1909.

17 The barn swallow (*hirundo rustica*) was 'appointed' to the position of Estonian national bird in 1962, at the peak of socialist optimism in the Soviet Union when the authorities launched campaigns to invent national symbols. The barn swallow, like a few other national symbols of the Soviet era, has not lost its meaning, its silhouette today marking locally produced food products, for example.

18 Maas, Winy; Andres Alver (2006). Muuseum ei pea viitama suitsupääsukesele. [Museum needn't referre to barn swallow.] *Eesti Päevaleht* 17.1.2006, http://www.epl.ee/?artikkel=309709, Reader's comments on the article.

The architectural competition had effectively disrupted the process of authorship which had begun with popular opinion and the reawakening of a nation seeking to connect to its pre-Second World War past. The successful architects saw the project in terms of prestige – giving Estonia a landmark cultural building that could sit beside those in other European nations. Naturally, these architects thought spatially and sought to highlight the spatial authorship over which they had control. Although they had no expertise as historians, curators or Estonians, they aimed to find a symbol to open a contested issue of the recent history of the nation and give the control back to the people through spatial means. For the general public, the restitution of the nation concerned issues of historical memory. Space was a subordinate matter. Debate in the media reflected the positions of both these parties, but it did not create debate between them or result in public participation. Between these two groups sat the museum's curators who were primarily interested in the internal spaces of the museum and their authorship. For them, restitution was a point of observation rather than participation and their interests focused on the academic understanding of folk culture as an important underpinning of national identity.

2. Participation

What emerges as critical to the production of the new museum is this distinction between producer (author) and consumer (audience); in what ways, was the audience being engaged in the production process and did this constitute participation in the act of authorship? In early discussions, it appears that the public only had a voice in the various media and not in the project itself. This permitted this potential audience the role of commentator on a predetermined end product using their resources of historical memory and its fixed meanings and leisure consumption as a predesigned experience. The disappointment in the grand narrative of participation was itself a product of Estonia's recent cultural transformation; the approach simply had not found a foothold or become established as a social norm (Kalmus et al., 2009; Runnel et al., 2009). Discussing such issues with the public would have been unthinkable in the Soviet period and the rebirth of cultural participation has been slow after the crisis in early nineties, after the mass movements of independence had met the harsh realities of new state. But even in the West, architects are known for their possession and control of landmark projects. The new Estonian museum had become, first and foremost, a piece of architecture (Figure 3).

In discussions between the architects and the jury, the audience remained a vaguely conceptualised entity. It was to be the *target* of meaning-making and strongly guided rather by the architectural idea than by audience study. The jury felt they were delivering what they considered, the audience needed, indeed what

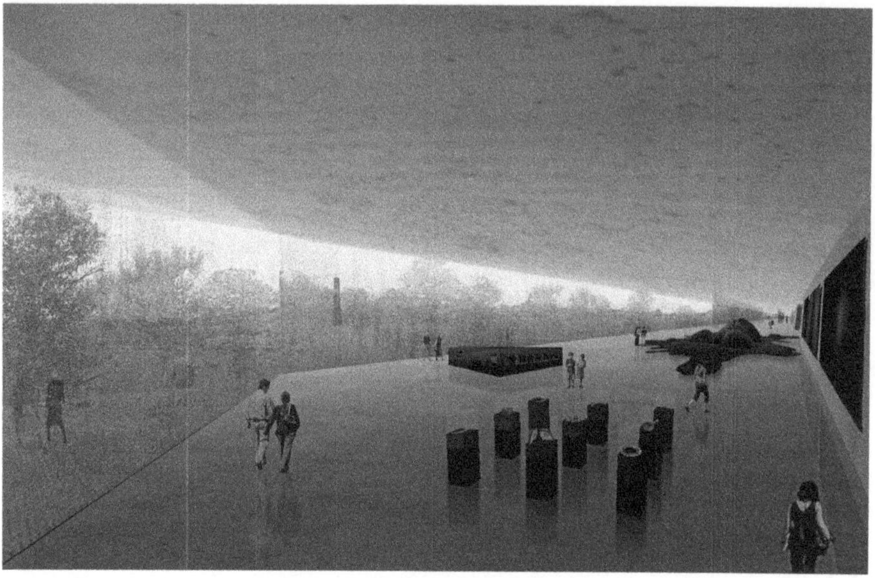

Figure 3: Visualisation of the new Estonian National Museum in the winning architectural project. Photo: Dorell. Ghotmeh. Tane / Architects

the nation needed: something what could have been considered progressive and 'trendy'. The member of the architectural jury Winy Maas stated:

> As a museum of national heritage, displaying the life and culture of the nation in its temporal, spatial, and social diversity, it complements and at the same time competes with emerging pop culture as expressed through the media, fashion, travel and lifestyle. Therefore, the committee was looking for a design that would give the new museum an active role in the new global pop culture that transcends national boundaries, and that would also transform the rather passive, 'dusty' attitude towards the museum into an active, 'hip' presence, that would attract the younger generation and be functional and competitive in the international context.[19]

As this quote highlights, the audience was kept at a distance – socially and conceptually – and remained as abstractly and taxonomically conceived in this process of conceptualisation as in the architects' and engineers' plans for the museum's public services and spatial logistics, which divided visitors up into general audience, those with special needs and those classified as children. In this preliminary

19 Paulus, Karin (2006). Mälu lahinguväljad. Intervjuu Winy Maasiga. [Memory debates. Interview with Winy Maas.] *Eesti Ekspress*. Available at: http://paber.ekspress.ee/viewdoc/82D7129DB71C5B0CC22570F8005ECDFA (accessed February 27, 2009).

phase, meetings between engineers, museum employees and architects never involved discussion of the audience or its potential involvement and it never appeared as a topic on meeting agendas.

The Director of the Estonian National Museum, Krista Aru, expected the new museum to convey a story that "should not be a dull monologue, but a lively dialogue, inspiring participation, research, discourse and continued study. A dialogue that attracts people, sparks ideas and encourages new endeavours" (Aru, 2006: 9). These words indicate a desire for the museum to be open, inclusive and engaging but there is no indication of how such a dialogue would be produced. As in museum culture generally, this Estonian director had the power to shape the institutional outlook (Tali, Pierantoni, 2008b), but all such directors work within cultures possessing established professional outlooks and long-term careers which can be resistant to change.

Some of the museum's curators and project managers certainly considered audience dialogue a new and challenging development. These curators retained an expert relationship with their audience which puts them in the position of authorship with regard to the museum's various messages. Curatorial interaction with the public follows formats established in the early years of the museum: distributing questionnaires for the museum's Network of Correspondents[20] in order to collect ethnographic information; conducting ethnological fieldwork for research and preparing temporary exhibitions, etc. This kind of audience engagement, despite containing dialogical elements, regards people rather as sources of information and authenticity for the museum, while not leaving much space for active interpretation of the audiences or shared authorship.

Perception of the needs of the audience by those controlling the project were mainly based on what they had seen in museums around the world, together with their own particular consumption preferences and cultural tastes. Thus land use, politics and archaeology were deemed 'boring', while volunteer fire-fighters during the first period of independence and the gendered division of everyday life were seen as 'sexy', in terms of audience appeal. When dealing with audience-related matters, staff attention centred on target groups with whom the museum had long had contact. Family members of curatorial staff also appear in conversations and were used to represent the public more generally.

---

20 The correspondents' network of the Estonian National Museum was founded in 1931 in order to collect extra data about the objects, collected to the museum. Today the correspondents' network comprises about 700 contributors both from Estonia and abroad, answering the questionnaires regarding both modernity and the recent past. The collection of materials on the topics chosen by the correspondents also continues.

These passing references to the museum's audience, however, did nothing to address the 55 per cent of people living in Estonia who had never heard of the Estonian National Museum.[21] Among this number, the Russian-speaking minority in Estonia is disproportionately significant. The presence of other nationalities in Estonia was discussed with regard to the future museum but they were not been integrated into the project. The draft of the profitability and feasibility analyses of the future museum declared that future exhibition spaces would tell the story of non-Estonian groups, particularly Russian-speakers, for the first time in the nation's history. The execution of this agenda for minority engagement seems an obvious moment to introduce shared authorship but it would need to overcome the established discourse of ethnic nationalism and the museum's passive engagement with the audience. The ethnographic nature of the Estonian National Museum positioned the audience as a subject and source of information; its marketing efforts saw them only as notional 'target-groups'. Perhaps for the museum to realize its potential, change was required not just internally but in the audience itself which had to learn to become active in civic and cultural participation.[22]

3. Towards innovation and participation

The word 'nation' in the name of the Estonian National Museum indicates that this is an institution serving a living population and a living culture facing the turmoil of today while preparing itself for tomorrow. It is not a word that merely signals that it possesses the residues of earlier or vanished manifestations of the nation. It indicates that the museum has a role in opening up and supporting change, and particularly in addressing an Estonia that is multicultural and situated in a modern global environment. It is in the phase of production, that brings into being the new museum, that such reconfiguration can take place but in the case of the Estonian National Museum this phase has produced two quite separate discourses which have failed to communicate. One, popular, vernacular and traditional, is possessed by the public; the other, intellectual, academic and architecturally postmodern, is in the possession of professionals.

In its current form, the Estonian National Museum remains home to the first discourse: an explanatory, fixed view of history based on essentialist views of culture. Although developed by museum curators who researched their interpretations, the exhibition nevertheless communicates a commonsense discourse so

21 Turu-Uuringute AS (2008). Eesti Rahva Muuseum 2008. Elanikkonnaküsitluse tulemused. [Estonian National Museum 2008. Results of representative survey.]

22 Morrone, Adolfo (December 2006). Guidelines for Measuring Cultural Participation. UNESCO Institute for Statistics. Available at: http://www.uis.unesco.org/template/pdf/cscl/framework/CUL_particip.pdf.

well established in Estonian society that one might consider it the national canon. It perceives culture as a bounded unity – something to be secured and protected, using the physical evidence of material objects – which helps express a wider distinction between 'them' and 'us'. In 1994, when the current exhibition was opened, this seemed the correct approach as the country had only recently regained independence and needed to rapidly recover its identity in a shared sense of belonging. As such it reflects popular conceptions of national culture which, dominated by cultural anxieties, sought security in a timeless romantic peasant past. If the museum was to reinvent itself with this discourse, it would remain centred on a depository for as one newspaper remarked, "where the treasury of the nation is located, is also its mentality".[23] This process of collection building would result in ritualised confirmation of (ethnic) nationhood and national identity. From the debate surrounding the new museum, it is clear that the Estonian public would consume such an exhibition as authentic.

The professionals involved in shaping the new museum, however, have become attached to a discourse which is intellectual, academic and postmodern (see, for example, Bhabha, 1994; Hann, 1994; Baumann, 1996). It favours negotiation and the collection becomes rather less central to the museum's purpose. This vision privileges the museum as a particular kind of communication institution, a place that potentially enables changes in what we know and how we think about things, a place that influences attitudes and becomes a laboratory of value systems and identities.

The ethnology curators who are to create the new permanent exhibitions exist between the public and the leaders of the architectural and conceptual project for the museum as a whole. Because of the expectations of the general public, they will, however, be attached to some extent also in the future to a view of culture that is static, fixed, objective, consensual and uniformly shared by the majority of the (ethnic) community. This is seen, for example, in the folk art consultation centre, to be opened in the new building, which would draw its knowledge from the museum collections and give the public advice concerning the 'rights and wrongs' of how to wear ethnic dress, amongst other things.

The missing ingredient at the heart of these issues is the audience itself. Without dialogue during the preparation of the museum, a lot depends on guesswork – or rather, the preferences of those in control on the basis of their professional experience and greater exposure to museums as creative and evolving institutions. Of course, the team developing the museum may have genuine concerns that such consultation will result in compromise and the diminution of concept. The only

23 Mikelsaar, Raik-Hiio (2003). Rajame Raadile estoloogiakeskuse ja mini-Eesti? *Tartu Postimees* 98 (1438), 23.5.2003, p. 2.

time the audience was more fully considered was during the writing of the EU application to secure funding for the new museum building. But even then the project-writing company undertaking this work were interested only in market targets and cost-benefit – the audience as numbers rather than people.

A dialogical national museum permitting shared authorship calls not only for the modernisation of the museum's communication and consultation apparatus, it may also require a fundamental shift in the underlying concept of 'Estonianness'. Without these changes the museum cannot be reinvented but rather would continue to perpetuate itself in its own values. Yet, audience awareness shows that this does not reflect the modern Estonian state, the realities of the past or the nation's modern context. The key issue in the transformation process, or 'reinvention' of the Estonian National Museum, is whether the museum will be able to substitute or enrich ethnic nationalism with civic nationalism. This could be done by supporting and fostering participation; offering audiences an opportunity to engage in re-writing stories about the Estonian past. This would also enable the Estonian National Museum to broaden the default concept of Estonian identity. The role of the museum is, in this setting, to provide *all* audiences with thought-provoking materials rather than just comfort blankets. If the national museum is able to take on this active role in re-imagining the nation in this way, then rather than simply being a site of social memory, it can also be a place of cultural innovation and cohesion.

Acknowledgements

The publishing of this article was supported by the Estonian Science Foundation grant no 8006 and the Estonian Ministry of Education and Science target financing project no SF0180002s07.

References

Aljas, Agnes (ed.) (2006). *Eesti Rahva Muuseumi avalik rahvusvaheline arhitektuurivõistlus. Kataloog. Estonian National Museum Open International Architecture Competition: Catalogue*. Tartu: Estonian National Museum.

Aronsson, Peter; Andreas Nyblom (eds.) (2008). *Comparing: National Museums, Territories, Nation-Building and Change: NaMu IV, Linköping University, Norrköping, Sweden, 18–20 February 2008: Conference Proceedings*. Linköping: Linköping University Electronic Press.

Aru, Krista (2006). The New Estonian National Museum Building – an Opportunity. In Agnes Aljas (ed.). *Eesti Rahva Muuseumi avalik rahvusvaheline arhitektuurivõistlus. Kataloog. Estonian National Museum Open International Architecture Competition: Catalogue*. Tartu: Estonian National Museum, p. 9.

Baumann, Gerd (1996). *Contesting Culture: Discourses of Identity in Multi-ethnic London.* Cambridge: Cambridge University Press.

Bhabha, Homi K. (1994). *The Location of Culture.* London: Routledge.

Black, Graham (2005). *The Engaging Museum: Developing Museums for Visitor Involvement.* London: Routledge.

Bourdieu, Pierre (1993). *The Field of Cultural Production: Essays on Art and Literature*. New York: Columbia University Press.

Carpentier, Nico (2007). Participation and Interactivity: Changing Perspectives. The Construction of an Integrated Model on Access, Interaction and Participation. In Virginia Nightingale, Tim Dwyer (eds.). *New Media Worlds: Challenges for Convergence.* Oxford: Oxford University Press, pp. 214–230.

ENM (2006). General Committee's Remarks Concerning the Winning Entries. In Agnes Aljas (ed). *Eesti Rahva Muuseumi avalik rahvusvaheline arhitektuurivõistlus. Kataloog. Estonian National Museum Open International Architecture Competition: Catalogue.* Tartu: Estonian National Museum, p. 25.

Hann, Christopher (1994). *Social Anthropology*. Lincolnwood: NTC Publishing Group.

Kalmus, Veronika; Pille Pruulmann-Vengerfeldt, Pille Runnel, Andra Siibak (2009). Mapping the Terrain of "Generation C": Places and Practices of online Content Creation among Estonian Teenagers. *Journal of Computer-Mediated Communication*, 14 (4): 1257–1282.

Kennedy, Michael D. (2002). *Cultural Formations of Post-Communism: Emancipation, Transition, Nation, and War.* Minneapolis: University of Minnesota Press.

Lauristin, Marju; Peeter Vihalemm (2009). Internal and External Factors Influencing Estonian Public Agenda during Two Decades of Post-communist Transformation. *Journal of Baltic Studies*, 40 (1): 1–28.

Maas, Winy (2006). The Final Summary by the General Committee – 16 January 2006. In Agnes Aljas (ed.). *Eesti Rahva Muuseumi avalik rahvusvaheline arhitektuurivõistlus. Kataloog. Estonian National Museum Open International Architecture Competition: Catalogue*. Tartu: Estonian National Museum, p. 23.

van Mensch, Peter (2005). Annotating the Environment. Heritage and New Technologies. *Nordisk Museologi,* 2005 (2): 17–27.

Peterson, Mark Allan (2003). *Anthropology and Mass Communication: Media and Myth in the New Millennium*. New York: Berghahn Books.

Runnel, Pille; Pille Pruulmann-Vengerfeldt, Kristina Reinsalu (2009). Estonian Tiger Leap from Post-communism to the Information Society: from Policy to Practices. *Journal of Baltic Studies*, 40 (1): 29–51.

Stead, Naomi (2004). The Semblance of Populism: National Museum of Australia. *The Journal of Architecture,* 9 (3): 385–396.

Tali, Margaret; Laura Pierantoni (2008a). The Processes of Contemporary Museum Constructions: Designing Public Space and Engaging Audiences. In Peter Aronsson, Andreas Nyblom (eds.). *Comparing: National Museums, Territories, Nation-Building and Change: NaMu IV, Linköping University, Norrköping, Sweden, 18–20 February 2008: Conference Proceedings*. Linköping: Linköping University Electronic Press.

Tali, Margaret; Laura Pierantoni (2008b). Ambitions and Realities: Developing Exhibition Programs in New Museums. NaMu Conference: European National Museums Encountering a Globalized Culture. (18 November). Presentation, Oslo.

When the Museum Becomes the Message for Participating Audiences

Pille Pruulmann-Vengerfeldt

Pille Runnel

1. Introduction

The past twenty and more years have been characterised by several significant transitions in society. The ongoing democratic revolution (Mouffe, 2000), intensified by the end of the cold war, the collapse of the Soviet Union and the consequent re-shaping of Europe, the constant discoveries in the area of human biology (and especially genetics), the increased relevance of information and communication technologies such as computers, mobile telephones and the internet are just a few of the more remarkable ones. These processes have also brought forward dependency on technology and increased the perception of risk and uncertainty in society (Beck, 2005).

The development and spread of the many variations of the democratic worldview along with new technological facilities has also affected museums, influencing them to become more communicative. Two core processes in museums, digitisation and democratisation, lead museums to focus on the dialogue with its audiences – providing more information is no longer considered sufficient.

The increase of communication and dialogue in museums has several consequences. On the one hand, the vast resources of cultural heritage can and are being made available through digital technologies. On the other hand, the dialogue at the museum level is much broader and has to be seen as part of the general democratisation of society. Democratising knowledge institutions such as museums helps society to come to grips with the pressures caused by general ambiguities in society by providing access to interpretations rather than ready-made solutions.

Museums, which have traditionally been institutions of knowledge and truth (albeit to varying degrees), are experiencing the need to open their collections, exhibitions and educational work in order to better fulfil their role as a public institutions within the democratic framework. One way of doing this is by increasing participatory activities within the museum environment, which will be the focus of this article.

Participation is often linked to the concept of interactivity in museums (e.g. Barry, 1998). Indeed, being engaging and interactive, especially through new technologies, is becoming increasingly the focus of museum work (Ciolfi, Scott,

Barbieri, 2011). However, this article takes a step further and argues that interaction and engagement are not enough in themselves. Although we discuss interactivity here in passing, we will not focus on this theme. Even if the concept is quite familiar for museums – especially in connection with new technologies – interactivity is generally not used to consciously facilitate democratic participation in the museum context. Rather it is 'just' a potential tool for engagement, which in reality more often offers support to the educational framework according to which interactive elements in museums are approached as learning tools.

Thus, while within the museum world interaction has the concept of pedagogy as its focus, participation is understood in the context of this article as mutually beneficial, respectful and to a certain extent, aiming for balanced power relations, or at least acknowledging the worth of discussion partners. Through this emphasis on respect and partnership, social interaction and participation become located at another, more fundamental, level of democratic support. In this article, we shy away from the minimalist approach to democracy, which would limit it to institutionalised politics. Instead, we take a more maximalist approach and look at the democratisation of society at large, acknowledging the importance of a well functioning civil society, thus extending the notion of citizenship beyond institutionalised politics.

The concept of 'participation' originally signified the cooperation of institutions and either the community or individuals, although as it has become used more widely, it has lost quite a lot of its meaning. Already in 1970, Carol Pateman (1970: 1) notes that "any precise, meaningful content has almost disappeared" from the term participation. The democratic-theoretical understanding of participation still has its dominance, but in this article our ambition is to extend this notion to museums, in order to understand participation in relation to the variety of roles outlined above. Peter Dahlgren (2006: 24) helps with the clarification of some key terms: "Engagement generally refers to subject states […] mobilised, focused attention." He sees engagement as a prerequisite for participation, as the latter would be "connecting with practical, do-able situations, where citizens can feel empowered […] it involves in some sense 'activity'". For Dahlgren (2006), although both participation and engagement are anchored in individual, they do have important collective dimension as they imply being connected to others via civic bonds.

In her book, *The Participatory Museum*, Simon (2010) argues that with museum participation, the key is finding out what function participation supports. In contrast to many ladder-based approaches towards participation (Arnstein, 1969; OECD, 2001; IAP2, 2007; Pruulmann-Vengerfeldt, 2010), Simon indicates that in the context of museums, different approaches to participation are better understood as a matrix in which in some of these instances the role of the museum is greater, while in some other cases the role of the museum decreases and leaves more control with audiences (Table 1). Simon (2010) stresses that it would be wrong to approach any of these participatory ideas as hierarchical, but rather these options are complementary and depend on museum's aims and possibilities. As Mariana Salgado (2009) argues, this does not imply that the traditional museum institution has disappeared, despite the shift of museums from being collection-centred towards being visitor-centred. However, she also sees this shift as the key to museums becoming participation-friendly institutions. McLean (2007) argues that this shift occurred when participation was understood to be part of learning, which differentiated this phase from earlier initiatives in which people are involved in museum activities either through collecting, commenting or interpreting. Thus, in many instances, participation and engagement become seen as either prerequisites or additions to fulfilling various museum roles.

Table 1: Different museum participation possibilities, adapted from Simon (2010)

	Contributory	**Collaborative**	**Co-creative**	**Hosted**
Control over the agenda and over the outcome	Museum	Museum more than participants	Equal/participants more than museum	Participants (with rules and some limitations from the institution)
Number of participants and their commitment	Potentially very many, but limited or no commitment	Smaller numbers, some casual joiners, but most with intention to participate, thus relatively small numbers	Relatively small groups, committed through the whole process	Relatively small groups, who need additional support for their own project
Participants interaction	Individual interacts with the content of the museum and possibly with other participants contributions	Individual interacts with content and institution and possibly with other participants contributions	Success presumes interaction with institution and other participants and co-operation	Success relies on good interaction with other participants forming a community or network
Goals for how non-participating visitors will perceive the project	Visitors see themselves as potential participants and see the institution as interested in their active involvement	Visitors see the institution as a place dedicated to supporting and connecting with community	Visitors see the institution as a community-driven place. It will also bring in new audiences connected to the participants	The project will attract new audiences who might not see the institution as a comfortable or appealing place for them

In the following, we will firstly give a short overview of museum history and introduce different positions the museum can have towards its audiences from the historical perspective. This will help to ground the discussion of the participation in the overall development of the museums as public institutions. This overview will provide insights into how the often conflicting approaches towards museum work have evolved over time and are still in the process of change. Secondly, different perspectives towards audiences will be mirrored in the discussion of three intersecting fields (social, cultural and economic) that museums operate in. In the third section, we will use core questions from the classic communication transmission model (Lasswell, 1948; McQuail, Windahl, 1993), with a twist on participatory communication focusing on the dialogue between the museum and its audiences. We will discuss the issues of museum participation through the lens of museums, by looking at which roles museums take in audience communication, why museums need to make people more aware of participation and what position is assigned to the participants and audiences in these participatory processes.

Our concern is not with audience motivations and what they gain from participating in public institutions. Rather, we take to a certain extent the normative position that institutions need to support participation. We assume that by looking at these different roles and areas where museums operate, we can better understand and support institutional motivations. Many of the discussions outlined here, centring on the museum institution, could be extended to other public institutions, which are opening themselves towards public participation. This article will hopefully contribute to a larger debate on the changing roles of public knowledge institutions in contemporary society.

2. A short and non-comprehensive history of museums

The changing roles of the museum can be exemplified by briefly looking at museum history. As Hooper-Greenhill (1995) explains, the stories of the museum's past are complex and illustrate many conflicting developments. Early museums were cabinets of curiosities with public access for the 'respectable' as early as 530 BCE (McDonald, 2006). In this kind of museum, the owner and his staff opened the doors and displayed the collection for the selected few. Audiences for this kind of institution were relatively closed groups and the communicative potential of this kind of museum was more related to influence and affluence than to knowledge and education. Museums became public institutions only during the Renaissance. This brought the development of a variety of functions, including socialising and educational aspects, collecting and also preserving and displaying the collections. The functions in the public institutions evolved, resulting in increasing complexity within the museum institutions themselves. Different functionalities of the museum became separated in different departments and thus distanced from each other.

This changed again in the second half of the 20th century when contemporary museums developed an increased coherence in relation to its various functionalities, represented by everyday cooperation at the organisational levels and by the overlapping and co-occurring of various processes. Museologist Peter van Mensch justifies this change by suggesting that today's museum needs to overcome these departmental differences in order to start thinking in terms of the visitors to whom the services of the museum are oriented (2005).

This was not the only change for museums have been investigating notions of 'ecomuseum' or community museum (de Varine, 1998), 'dialogic museum' (Tchen, 1992) and paid attention to the changing relations between museums and communities (Karp, 1992) for over forty years (Pollock, 2007). Thus, they became implicated in what Giddens (1998) labelled the responsibility of public institutions to contribute to the democratisation of democracy. Public knowledge institutions, such as museums, need to become what van Mensch (2005) calls laboratories and meeting points for discussions and new initiatives. In other words the 'sanctum-museum' needs to become a 'laboratory-museum' (Mairesse, 2003), respectful of the expertise of the museum staff and its experts, but at the same time open to a continuous dialogue with the outside worlds that sometimes come to visit it. More specifically, a 21st century democratic and reflexive society needs museums that encourage society's publics to attribute meaning to the cultural objects that are on display (Hein, 2006).

At the same time, museums, together with many other institutions face the challenge of competing for people's time. Entertainment and leisure seem to be universally acknowledged ways of organising this. One common way to achieve attention from audiences is the celebrification of museum objects. Rojek (2001) defines celebrification as the attribution of glamorous or notorious status to an individual within the public sphere, although the concept can also be used for objects. Celebrification occurs in many arenas, and museums promote certain objects in their collection to the celebrity status in the hope of gaining more attention (and visitors). Van Mensch's (idea of a) museum is an institution that is very close to its audience; it can be said that the museum institution, hoping to gain visibility and connection with its audiences through the celebrification process in fact distances itself from its audiences by making them consumers-worshipers of glorious collections.

These above-mentioned processes occur simultaneously in the contemporary museum: the organisational division of labour (which has become more porous), the celebration of partnerships, and the glorification of objects. This also implies that in different museums, the attention for the audiences and their ways of dealing with the visitor differs. These also impact on the ways that museum institutions allow or disallow participation. In order to capture these diverse and overlapping practices, three fields are introduced, within which these practices are embedded: the cultural field, the economic field and the public field.

3. Museums in their contesting and intersecting fields

The notion of fields is borrowed from Bourdieu's idea (1998) that different fields carry different operational logics. The framework of fields helps to explain some of the contradictory and overlapping social processes museums seem to undergo. Museums operate on three key fields – cultural, economic and political, fulfilling three key institutional roles: being simultaneously a cultural, public and economic institution (Figure 1). The related roles, responsibilities and needs are often conflicting. Some of these role changes are emerging alongside the changes outlined in museum history, but as outlined in the discussion about museum history, none of the previous roles has completely disappeared. At the same time, the redefinition of the museum is on the agenda, and museum culture in general is seen in need of reorganisation (Imminga, 2010: 9). Our concerns are then how these different aspects relate to public participation and how they provide reasoning, motivation and support for participation.

As a *cultural institution*, museum roles include preserving, collecting, interpreting and mediating heritage to publics. As a *public institution*, museums are socialising and democratising agents and thus share the role of educational institutions. The third role comes from the museum as an institution operating within the *economic field*, where museums need to compete in the open market for clients' leisure and free time. Here museums need to collect revenues and attract visitors. Even if museums are publicly funded, there is an increasing pressure for additional revenue collecting. DiMaggio (1985) described – over 25 years ago and writing about the US – how museums face many contradictory demands and that they often operate in paradoxical situations in which they are publicly funded and expected to produce public good and be 'non-profit', while also being expected to compete in the free market. Falk (2009) also places all leisure activities at the same level and describes how for the people, museums are just another place to go. At the same time museums today are increasingly seen as vital parts of the creative economy and their roles and functions are being acknowledged as actively negotiated and fluid. Lord (2007: 8) makes a similar argument when he writes that in order to benefit from the creative economy, museums need to be dialogic and truly open to diversity and interdisciplinary approaches and they could become

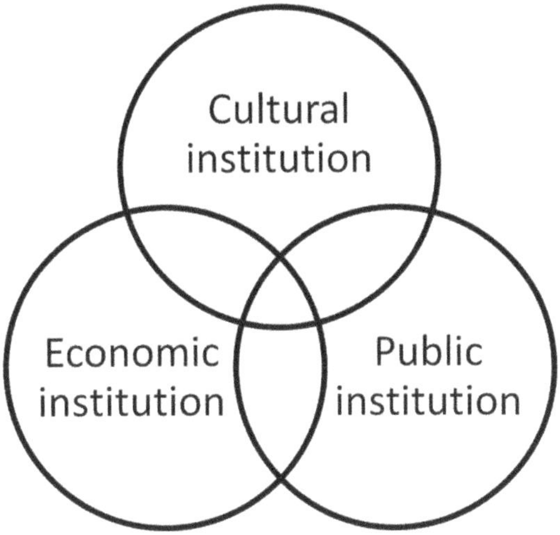

Figure 1: Key domains of the contemporary museum

cultural accelerators, forums and sites for debates. Otherwise, they might benefit in the cultural economy only through cultural tourism.

The roles stemming from different fields also have commonalities and overlaps with each other; often the goals and means are shared. At the same time, there are still plenty of other cases where the roles can be conflicting, causing tensions within the museum and between the museum and its communities. In many cases, the interpretations of these institutional roles depend on professional museum workers as well as on their publics. Negotiation of the functions sometimes occurs in peaceful dialogue, whereas in other instances these roles can be sources of intense conflicts either within the museum or between museum and its many stakeholders. Elsewhere, we have discussed some of these conflicts regarding the perception of the roles of the museum in the context of the Estonian National Museum, where the conflicting roles are the interconnected views of architects, museum professionals and the general public (Runnel, Tatsi, Pruulmann-Vengerfeldt, 2010). Enabling and increasing participation in museums can be one way of overcoming the differences of opinions, but many of the expectations are also there to hinder the possibilities of participation.

4. Museum is a voice is a message is a medium

In this article, we look at the museum as a site of participation for different audiences through the lens of the classical communication model of Who? Says What? To Whom? (Laswell, 1948; McQuail, Windahl, 1993). Using this basic communication model helps to structure the elements of participation in the museum context. The focus of the analysis will be framed by the fact that museums operate in three key fields: cultural, economic and political. They thus carry three different but still co-existing and overlapping roles. The three fields, combined with the three topical questions will be used to discuss how museums can deal with increased societal expectations and needs to organise more (maximalist forms of) participation.

4.1. The museum as a communicator – positioning "Who?"

The structure of this part of the article follows the logic of the three fields – cultural, economic and political – inspired by Bell (1976) and Bourdieu (1998). However, the idea behind using these three fields (and they by no means cover all the activities of a contemporary museum) is to distinguish between the different operational logics of the different areas. In many instances the different fields can be either more or less dominating for a particular museum.

If the museum looks at audience participation from the position of the *cultural institution*, then the role of the museum in inviting people to participate may very much depend on the types and identities of the museum. Although one can argue that museums and other knowledge institutions like libraries and archives have much more in common than often assumed, then in some of these instances distinguishing between an ethnographic museum, a history museum, an art museum, a children's museum, science museums, etc. may be justified. The issue here is that the museum as a cultural institution may have different possibilities and different reasons to invite people to participate. Potential reasons for this cultural institution perspective are the possibility to have visitors add artifacts or stories to the collections, the opportunity to make more engaging exhibitions that are enriched by visitor input, and to involve the visitor in a process of joint cultural production. There are also limits imposed upon participation, as museum workers define this process of cultural construction as the exclusive area of their expertise (Carpentier, 2011).

As an *economic institution*, the driving force for the museum would be making money/profit, and that would also be the key motivation for inviting people to participate, if museums decided to do so. Potentially, the cost of organising participation may be deemed too high. However, there might be different mechanisms by which participation would support the aim of money-making. It can be that participation helps to engage and attract visitors and make it more appealing to come to the museum and thus support marketing messages. It may be that with participatory activities, museums keep people longer on their premises and can profit from selling them refreshments. It can also be that participatory activities enable museums to add valuable items to the collections, making the museum generally more attractive. If carefully planned, participation and community involvement may also become important monetary resource through either helping to raise money for a common cause or by helping the museum to save money by outsourcing some of the activities to the community.

Museums as *public institutions* see their participatory role primarily through the need to empower people through participation. Here, civic engagement with the institution might mean that people leave the institution more knowledgeable, with a successful experience, with a sense of value and self esteem (coming from the fact that a knowledge institution finds individual contributions valuable). The added meanings of participation might come from the interaction with experts, whereas in other instances it is the message from the museum saying that people outside museums carry some kind of valuable expertise the museum needs. Again, this role could potentially work against participation, as museums might decide to stick to the more traditional informational and educational definition of the public institution.

4.2. Participating in what?

In the introductory part of this article, we referred to the overarching aim of the museum to invite its visitors and users to participate within a changing societal context. The different roles of the museum also mean that different aspects of participation are relevant to each of these roles.

The definition of participation as it is manifested in different fields is outlined at the next schema (Figure 2). Each domain in which the museum operates is described by its distinctive understanding of participation and user engagement. For each field the meaning and aim of participation differs. In each particular field the notion and understanding of participation is brought into the museum using the concepts and reasoning of those particular fields. Thus in order to understand museum participation, we need to analyse the field-based logic and motivations behind the participation. Borrowing from the ladders of participation approach (whilst maintaining a critical distance), we can distinguish more active and more passive relations to audiences.

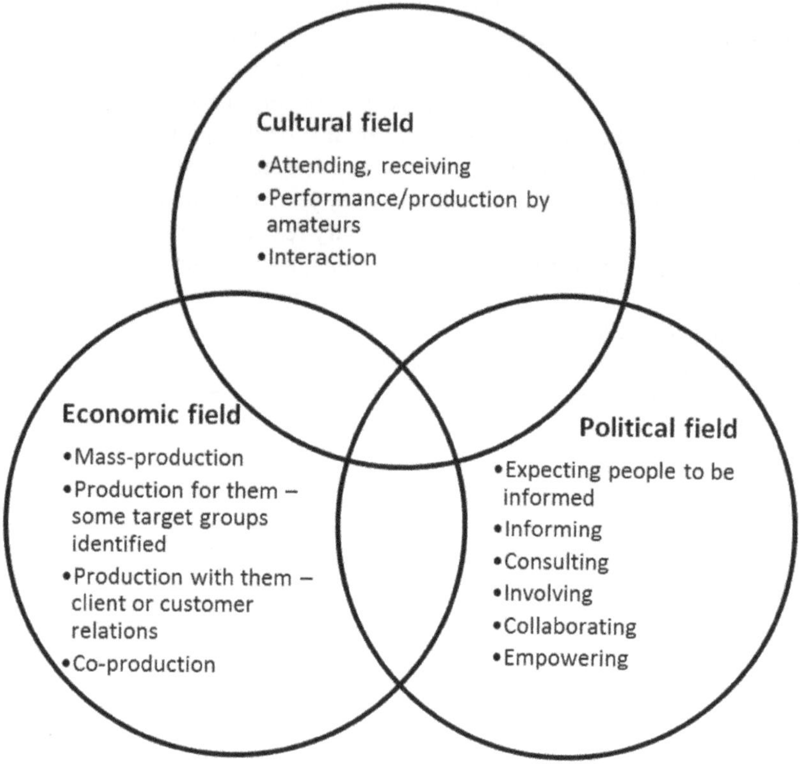

Figure 2: Participation and audience relationships in the different fields of museum operations

We should be careful not to blindly copy the active/passive approach, as it is not without its problems. In the context of the *cultural institution*, Morrone in UNESCO's *Guidelines for Cultural Participation* (2006: 6–7) claims that it is difficult and unwise to attempt to reduce cultural participation to an active/passive scale. He proposes a distinction of attending/receiving; performing/producing by amateurs; and interaction. For Morrone (2006: 7) interaction is a process "defined by continuous feedback of flow communication between external source and a receiving subject." With this kind of definition of interaction he attempts to quantify and explain the experiences enabled by new digital media, distinguishing interaction from attending, and defining receiving as a third and distinctly different way of cultural participation. Similarly to Simon (2010), Morrone does not see these activities as in any way hierarchical, but rather as a way to distinguish three different media through which participation can happen. Here the element of control and power is not at all prominent in distinguishing the three levels of participation. However, Morrone (2006) clearly distinguishes the professional and amateur aspects of culture and limits the understanding of cultural participation to the amateur only. This implies that in the cultural field, Morrone takes the stance that everyone is an active participant.

When moving to the next field, we can see that in *the economic discourse*, the term involvement is used, rather than participation. Participation here is more about attracting the public to be involved in the activities offered by the institution. Those who become involved, are sometimes termed 'prosumers'. This kind of relationship between the institution and its publics corresponds to the museum's increasing demand to be interactive. In many cases, interactivity is seen as adding technological solutions or elements such as buttons, screens and multi-media to the exhibitions. The problem is that this can lead to deceptive interactivity, where a person is given the sense that he or she has control over the process, whereas the control in fact is pre-determined by others (by technological tools and the intentions behind them).

The understanding of participation in the *economic role* of the museum remains rather vague. While we can definitely see discussions of audience participation in the debates on marketing and organisational communication, there is little evidence of the systematic classification of participation in the whole economic field. The discussion in marketing has for the past 20 years moved from pro-duct placement towards customer relations and dialogue (e.g. Christopher, Payne, Ballantyne, 1991), and the new web 2.0 technologies have only reinforced that trend (see, e.g., Godin, 2008). In Figure 2, we list a number of potential economic relations, which could be seen as co-existing and emerging depending on various external or internal factors. In the first instance, the institution does not care for the market other than for its purchasing power. In the second, some target groups are specified and production is carried out for them. The focus on the relationship with people is illustrated by the idea of paying careful attention to customer or client needs, understanding the selected target groups carefully and almost co-producing with them as a result. Lastly, economic relationships can evolve into the co-production through mutual cooperation and partnership in the production process. These stages are also distinguished by different levels of control and in a way this hierarchy mimics the IAP2 (2007) participation model in the economic field. However, while in the public field relinquishing control can be seen as part of the motivation (empowering individuals, the citizens, to take control), the economic field has different operational logics; here giving up control is not an option at all. In the economic field, the ultimate key seems to be in understanding the customer and proposing mutually beneficial partnerships in order to maintain economic dominance and gain profits.

At the same time, creative economy discussions envision the people in the active role of being engaged and interested, while museums become passive sites for their creative forces. Here, dialogue and participation happens within the community and the museum's role in these processes is yet to be understood.

When looking at *political-democratically motivated participation* in the museum, or the museum as a public sphere institution, it makes sense to talk about stakeholder engagement or mobilisation where the aim is often to rally the visitor or users to some course of action. Here museums can become sites of public campaigns. The more subtle role of democratising democracy means that museums as public institutions also have a responsibility to educate people not only about museum contents, but also about participation as such. Hence, it might be relevant to discuss the distinctions of different ladder of participation approaches (e.g. OECD, 2001) and stress that although informing is not necessarily participatory, museums can and often do see civic education as part of their public role; and informing can become a prerequisite to mutually beneficial participation.

Political participation has probably been analysed and described the most thoroughly. In Figure 2, we have summarised the propositions of the International Association of Public Participation (IAP2, 2007) in order to approach political participation as providing information, consultation, involvement, collaboration and empowerment. These levels have a clear hierarchical structure. While each level is perceived as valuable, fulfilling specific goals, with its own specific instruments, the level of public impact is seen to be increasing with each subsequent stage. In the context of knowledge institutions, an additional level is described in this scheme: expectation that the public will be informed. This layer contains an expectation of a public institution that although the role and responsibility of an institution is to serve the public, the responsibility of looking for this public service is solely that of the recipient. This corresponds well to the traditional role of museum as collecting and preserving, where the value and quality of the collections are seen as important for future (potential) researchers and viewers as today's active citizens. This idea of maintaining collections for the future as the paramount role of the museum is in a way part of the museum viewed as a public institution that excludes in the present everyone – except professionals (and possibly the donators) – from its activities.

4.3. Naming thy partner – to whom does the message go?

In the museum context, audiences have a variety of names. While 'audience' comes from the field of communication studies, museums have also conceptualised the people on their premises. For instance, Peacock and Brownbill (2007) bring together concepts of 'audiences', 'users', 'visitors' and 'customers' (originating from four different paradigms) in an attempt to understand the users of online and offline museum environments. The museums have been looking at their 'people' from the perspective of friends, visitors, clients, users, participants, while new technologies and new economic relations also expand on the notion of prosumers (Toffler, 1980) and produsers (Burns, 2006).

As naming has its power, the naming of the people who come to the institutions can also empower or marginalise people. When museums looked at their visitors as 'the respectable' or as 'friends', and showing off items of curiosity was central to their communication, a fairly limited imbalance of power was inscribed in the interaction. The holder of the collections was superior to the viewers in many ways, although s/he was still dependant on the visitor's approval. In the original museums, superiority might have stemmed from interest, monetary value or societal position. When museums became institutions, superiority was tied to expertise on preservation or knowledge of the items. In the shift towards a more participatory museum, it should be acknowledged that participation will never be all-inclusive and equally empowering. As discussed above, the variety of approaches enables different levels of audience participation. Nielsen (2006) has proposed a 1:9:90 rule, claiming that on average, in large scale multi-user communities, most participants do not participate at all. Participants can be divided into regular and active participants on the one hand, and into those who engage themselves from time to time on the other. In the museum context, this means that only some visitors can be potential participants in museum activities. When the modern laboratory-museum is looking for partners, they need to take into account the fact that, according to Simon (2010), participation has to be valuable for the institution, the participants, and also the 'lurkers'. Thus when we discuss participants, the museum, the actively engaged group of people and others all need to be satisfied and supported.

Here, again, the different fields raise different expectations regarding participants. As discussed above, *cultural participation* as defined by Morrone (2006) expects reception, participation through amateur production and interaction through new technologies. Moreover, the roles of the participants can also include those of informant, expert, contributor or creator of other kinds of content.

Operating in *the economic field* means that museum institutions have had to start better understanding their audiences. Through learning more about its target groups and customers for marketing purposes, museums also foster their participation in the other (cultural and political) fields. The economic field in most the cases defines customers or consumers in a fairly passive way. Here the customers are seen as a source of knowledge in terms of 'what they want'. When we look at the concept of creative industries, the understanding of museums in the economic field changes again. Here museums are seen as the site for active, engaged and critical individuals who are inspired by the museum for their cultural work. However, there is less focus on the museum taking an active role in these dialogues.

The role of the museum as *public institution* offers more possibilities and also raises more expectations. This role implies that active engagement can be situated on many different levels. For museums, people who see the museum as a resource, people who act as quality contributors, or people who are partners in collaborative projects are all important. Of course, we should keep in mind that it is impossible to have all functions of the museum realised through co-creative or hosted activities, as this would be too resource consuming for any institution.

Although contributing, and possibly also collaborating, can be individual, participation can also have a more social dimension when a group of individuals works together with an institution. Arguably, only a group of people or a community with mutual awareness and an existing network can be a partner to the institution with the potential capacity to share power. Museums can look at the participation as a possibility to foster the birth of such community or network. Simon (2010) proposes five stages of participation,[1] which range from 'me' (where an individual consumes content) to 'we' (where individuals engage with each other and the institution becomes a social place full of enriching and challenging encounters). The stages in between help to link the visitor to the content, and through the content also to other visitors. Simon (2010) sees these stages as progressive and proposes that for the stage 5 experience, the groundwork of the other four stages is needed. While today's museums focus mainly on stages 1 and 2, the incorporation of other stages makes the participation more valuable for both the individuals and institution. When critically examining the IAP2 participation model (2007), one can see that more public involvement becomes possible only when audiences start working together rather than remaining in a one-on-one interaction with the institution. In those instances, the institution also has more control over the agenda and outcome of the participation. Organised or networked communities have more chances to co-create or to work with the museum in a partnership, as the interaction is less dependent on

1 Simon terms this social participation, a term which does not receive too much prominence here in order to avoid confusion.

individual capabilities. Many of the more complex participatory initiatives demand more resources from the participants and networks or community groups are better able to fulfil these demands.

5. By way of conclusion

In this article, the classic model of communication of Who says What to Whom has been used, in combination with three societal fields to map audience participation in the world of museums. It is important to see that the different fields of operation generate different demands for museums and the praxis of participation depends very much on the situatedness in these particular fields. The museum has always been a medium for many different messages and through the logics of participation the wider circles of people are included as communicators. Traditionally, museums narrate the stories of their owners – either private or public – although through the organisation of these participatory practices, museums can take a step towards diversifying these voices. The collections and exhibitions need to be sites of discussions in order to foster the civic skills of the audience, but also to fulfil the expectations of the cultural economy.

It is important to understand that participation in museums needs to be understood through the diversity of approaches – often there are manifold choices to be made, and the increased number of active participants or contributors can mean that the contributions become more superficial, whereas collaboration or partnership can only occur with limited numbers of individuals. Again, this is a reason to place more emphasis on the organised or networked audience. Whatever participatory structure is preferred, as long as the repertoires of the participation are diverse, the participatory aims of the museum can be seen to be fulfilled.

This article has focused on museums as institutions in public ownership. We have not paid attention to privately owned museums and their particularities. However, it is clear that privately owned institutions face the same struggles and often their need for participation is even greater because of their necessity to raise funds and community support for their survival. The museums have been and will continue to be media for many messages and this article has hopefully contributed to understanding the many perspectives museums can take towards participation.

It is vital that museums understand that unless they open many of their functions to the public, they are not able to fulfil the obligations/expectations placed on them. We have spent little time on discussing the socialising functions of museums, although these can only be fulfilled if society sees the museum as a valuable resource and as part of its everyday activities. The experiences of participation improve when we look at the participants not as isolated individuals but as a collective, interrelated entity, and when we foster their interactions. Museums need to be sites for community building and networking.

In many ways, museums – as reflexive knowledge institutions – can play a leading role by introducing and socialising audiences to the ideas of participation. This also means that the traditional understanding of museums as sanctums of truthful memories needs to be abandoned, as the more post-modern society needs reflexive citizens. Reflexivity comes only with practice, when existing knowledges are questioned and analysed. Instead of providing visitors with ready-made and perfect answers, museums can use participation as a way to entice and support critical thinking. In this fashion museums have increasingly played a role in introducing literacy skills to the citizens of today.

It would be wrong to state that we have to invent new kinds of audience relations for the museum. In a way it would suffice simply to return to the initial understanding of museum audiences as friends, strengthened by the current understanding of audiences as partners in the experimental knowledge laboratories in order to construct the approach that we need to bring to museums.

Acknowledgements

The publishing of this article was supported by the Estonian Science Foundation grant no 8006 and the Estonian Ministry of Education and Science target financing project no SF0180002s07.

References

Arnstein, Sherry R. (1969). A Ladder of Citizen Participation. *Journal of the American Institute of Planners*. Vol. 35, No. 4: 216–224.

Barry, Andrew (1998). On Interactivity: Consumers, Citizens and Culture. In Macdonald, Sharon (ed.). *The Politics of Display: Museums, Science, Culture*. London & New York: Routledge, pp. 98–117.

Bell, Daniel (1976). *The Coming of Post-industrial Society: a Venture in Social Forecasting*. New York: Basic Books.

Beck, Ulrich (2005). *Riskiühiskond, teel uue modernsuse poole [Risk society: towards a new modernity]*. Tartu: Tartu Ülikooli Kirjastus.

Bourdieu, Pierre (1998). *Practical Reason: on the Theory of Action*. Cambridge: Polity Press.

Bruns, Axel (2006). Towards Produsage: Futures for User-led Content Production. In Fay Sudweeks, Herbert Hrachovec, Charles Ess (eds.). *Cultural Attitudes towards Technology and Communication 2006*. Murdoch: Murdoch University, pp. 275–284.

Cameron, Duncan F. (1971). The Museum, a Temple or the Forum. In Gail Anderson (ed.). *Reinventing the Museum: Historical and Contemporary Perspectives on the Paradigm Shift*. Oxford: Altamira Press, pp. 11–24.

Carpentier, Nico (2011). Encarando a morte do Autor. O trabalho identitário do profissional de cultura e as fantasias de controle [Facing the death of the Author. Cultural professional's identity work and the fantasies of control]. *MATRIZes*. Vol. 4, No. 2: 183–204. http://www.matrizes.usp.br/ojs/index.php/matrizes/issue/view/10 (accessed August 10, 2011).

Christopher, Martin; Adrian Payne, David Ballantyne (1991). *Relationship Marketing*. London: Heinemann.

Ciolfi, L.; K. Scott, S. Barbieri (eds.) (2011). *Rethinking Technology in Museums 2011: Emerging Experiences. Proceedings of the International Conference.* Limerick: University of Limerick.

Dahlgren, Peter (2006). Civic Participation and Practices: Beyond 'Deliberative Democracy'. In Nico Carpentier et al. (eds.). *Researching Media, Democracy and Participation: The Intellectual Work of the 2006 European Media and Communication Doctoral Summer School*. Tartu: Tartu University Press, pp. 23–34.

DiMaggio, Paul J. (1985). When the Profit is Quality: Cultural Institutions in the Marketplace. *Museum News* Vol. 63, No. 5: 28–35.

Falk, John H. (2009). *Identity and the Museum Visitor Experience*. Walnut Creek, California: Left Coast Press.

Giddens, Anthony (1998). *The Third Way: The Renewal of Social Democracy*. Cambridge: Polity Press.

Godin, Seth (2008). *Tribes: We Need You to Lead Us*. New York: Potfolio.

Hein, Hilde (2006). *Public Art: Thinking Museums Differently.* Lanham, MD: Rowman & Littlefield Publishers.

Hooper-Greenhill, Eilean (ed.) (1995). *Museum, Media, Message*. London: Routledge.

Imminga, Marjolein (2010). The Paradoxical Situation of Museums in the 'New Economy': 'Nederlands instituut voor Beeld en Geluid', Museum of the Future? Erasmus University Rotterdam. Available at: http://oaithesis.eur.nl/ir/repub/asset/8093/Imminga.pdf. (accessed August 10, 2011).

International Association for Public Participation (IAP2) (2007). IAP2 Spectrum of Public Participation. Availabe at: http://www.iap2.org/associations/4748/files/IAP2%20Spectrum_vertical.PDF. (accessed August 10, 2011).

Karp, Ivan (1992). Introduction: Museums and Communities: The Politics of Public Culture. In Ivan Karp, Christine Mullen Kreamer, Steven D. Lavine (eds.). *Museums and Communities: The Politics of Public Culture.* Washington: Smithsonian Institution Press, pp. 1–17.

Lasswell, Harold D. (1948). The Structure and Function of Communication in Society. In Lyman Bryson (ed.). *The Communication of Ideas.* New York: Harper, pp. 32–51.

Lord, Barry (2007). *Manual of Museum Learning*. Lanham: AltaMira Press.

Mairesse, François (2003). *Le musée, temple spectaculaire*. Lyon: PUL.

Macdonald, Sharon (2006). *Companion to Museum Studies*. New York & Oxford: Blackwell.

McLean, Kathleen (2007). Surviving in Two-way Traffic. In Kathleen McLean, Wendy Pollock (eds.). *Visitors' Voices in Museum Exhibitions*. Washington, DC: Association of Science-Technology Centres Incorporated, pp. 8–13.

McQuail, Denis; Sven Windahl (1993). *Communication Models.* London: Longman.

Mensch, Peter van (2005). Annotating the Environment. Heritage and New Technologies. *Nordisk Museologi* 2005 (2): 17–27.

Morrone, Adolfo (2006). Guidelines for Measuring Cultural Participation. UNESCO Institute for Statistics. Availabe at: http://www.uis.unesco.org/template/pdf/cscl/framework/CUL_particip.pdf (accessed August 10, 2011).

Mouffe, Chantal (2000). *The Democratic Paradox*. London: Verso.

Nielsen, Jakob (2006). Participation Inequality: Encouraging More Users to Contribute. *Jakob Nielsen's Alertbox.* Available at: http://www.useit.com/alertbox/participation_inequality.html (accessed August 10, 2011).

OECD (2001). Citizens as Partners: Information, Consultation and Public Participation in Policy-Making. PUMA: OECD. Available at: http://www.ecnl.org/dindocuments/214_OECD_Engaging%20Citizens%20in%20Policy-Making.pdf (accessed August 10, 2011).

Pateman, Carole (1970). *Participation and Democratic Theory.* Cambridge: Cambridge Univesity Press.

Peacock, Darren; Jonny Brownbill (2007). Audiences, Visitors, Users: Reconceptualising Users of Museum On-line Content and Services. In Jennifer Trant, David Bearman (eds.). *Museums and the Web 2007: Proceedings*. Toronto: Archives & Museum Informatics. Available at: http://www.archimuse.com/mw2007/papers/peacock/peacock.html (accessed August 10, 2011).

Pollock, Wendy (2007). Voice of the People. In Kathleen McLean, Wendy Pollock (eds.). *Visitors' Voices in Museum Exhibition*. Washington, DC: Association of Science-Technology Centres Incorporated, pp. 3–7.

Pruulmann-Vengerfeldt, Pille (2007). Participating in a Representative Democracy. Three Case Studies of Estonian Participatory Online Initiatives. In Nico Carpentier et al. (eds.). *Media Technologies and Democracy in an Enlarged Europe. The Intellectual Work of the 2007 European Media and Communication Doctoral Summer School.* Tartu: Tartu University Press, pp. 171–185.

Rojek, Chris (2001). *Celebrity.* London: Reaktion Books.

Runnel, Pille; Taavi Tatsi, Pille Pruulmann-Vengerfeldt (2010). Who Authors the Nation? The Debate Surrounding the Building of the New Estonian National Museum. In Simon J. Knell et al. (eds.). *National Museums. New Studies from around the World.* London and New York: Routledge, pp. 325–338.

Salgado, Mariana (2009). *Designing for an Open Museum, an Exploration of Content Creation and Sharing through Interactive Pieces*. Jyväskylä, Publication Series of the University of

Art and Design Helsinki A 98. Available at https://www.taik.fi/kirjakauppa/product_info.php?cPath=23&products_id=157 (accessed August 10, 2011).

Simon, Nina (2010). *The Participatory Museum.* Santa Cruz (Calif.): Museum 2.0.

Tchen, John Kuo Wei (1992). Creating a Dialogic Museum: The Chinatown History Museum Experiment. In Ivan Karp, Christine Mullen Kreamer, Steven D. Lavine (eds.). *Museums and Communities: The Politics of Public Culture.* Washington: Smithsonian Institution Press, pp. 285–326.

Toffler, Alvin (1980). *The Third Wave.* William Morrow.

Varine, Hugues de (1998). Rethinking the Museum Concept. In John Aage Gjestrum, Marc Maure (eds.). *Okomuseumsboka – identitet, okologi, deltakelse.* Tromso: ICOM Norway, pp. 33–40.

A Multi-Method and Multi-Site Interventionist Approach to Studying Audience Participation in Museums[1]

Pille Pruulmann-Vengerfeldt

Pille Runnel

Agnes Aljas

In recent years the relationship of museums and their audiences has been much discussed in relation to institutions role in widening participation and social inclusion, connected to technological innovations and social media developments. Eleanor Hooper-Greenhill has stated that the most important change of priorities in the museum field in last century has been the turn to the visitor. The turn comes mostly from economic impacts that museum and cultural field has had in recent decades. This turn has needed new skills of the museum workers, re-priotisation of resources and reconceptualisation of museums policies and plans (Hooper-Greenhill, 2011: 362).

Nowadays museum studies understand visitors beyond the classical site-visit situation and acknowledge that museum experience starts well before the visitor steps through the museum door. Museum visitors are not seen as mass public, but as individual interpreters with their own social contexts. This also calls for approaching museum publics as audiences rather than visitors. Museum institution does not distance itself from its audiences by making them consumers-worshipers of glorious collections, but hopes to gain visibility and connection with its audiences through dialogue, questioning and problematisation with subjects that matter to audience. Stylianou-Lambert (2010) sums up the developments in the field of museum visitor studies[2] since the 1990s, showing how this knowledge and these approaches have been taken into account in museums and museum studies and have led to a paradigm that presents the museum as an "open work that

1 An earlier version that is theoretically extended, and with different subset of empirical examples is published as: Pille Pruulmann-Vengerfeldt, Taavi Tatsi, Pille Runnel, Agnes Aljas (2014). Researching Audience Participation in Museums: A Multi-Method and Multi-Site Interventionist Approach. In Geoffroy Patriarche, Helena Bilandzic, Jakob L. Jensen, J. Jurišić (eds.). *Audience Research Methodologies Between Innovation and Consolidation.* New York; London: Routledge, pp. 87–106.

2 "Visitor studies" is an umbrella term for a range of different forms of research and evaluation involving museum and their actual, potential and virtual visitors which collectively might be termed the "audience" for museums (Hooper-Greenhill, 2011: 363).

is completed by the visitor" (Stylianou-Lambert, 2010: 137). However, there is also a tendency to "underestimate power issues, while romanticizing the power of audience activity, thereby ignoring issues of responsibility" (Stylianou-Lambert, 2010: 141).

In order to analyse museum audiences critically, we need to reconsider and, when necessary, to develop and apply the methodology for capturing the emerging social aspects in museum communication and take into account a more dialogical context. The changes in conceptualising museum audiences also show that in the visitor research, it is not sufficient to stay at the field of museum visitor studies, but this task, we argue, calls for applying analytical toolkits of both museology and the study of media audiences, where the concept of audience is better theorised and explored. New research methodologies should be considered, but also developed, combining knowledge from visitor research with that of the media audience studies, where different roles of audiences are being acknowledged and revealed and audiences are seen as consumers and producers – content-creators or co-curators of knowledge. Also, with the emergence of new technologies, participatory culture (Jenkins, 2006) and audience participation (Carpentier, 2011) are increasingly discussed and analysed in media studies.

In her book *Museums and the Public Sphere* Jennifer Barrett (2011) states that museums are highly visible institutions and their existence is often justified on the grounds of 'relevance' to the 'public', which is used loosely (Barrett, 2011: 1). There is still struggle of old rhetoric of 'public' and new practices and new types of space designed to attract new audiences (Barrett, 2011: 3) in context of new museology.[3] The ongoing quest to be attractive and relevant has been the concern for museums since the invention of public institution through the centuries (Dana, 2004: 1917). In late eighteen century Europe the tendency to conflate state and 'the public' becomes common – and is still ongoing, as institution of state (representative) and public (of the people) (Barrett, 2011: 6).

Currently museums are experimenting to strengthen the participation of the public in two-way conversations between the public and the institution, most often using either the internet, especially social media or exhibition spaces for facilitating participatory activities. Especially digitalised collections, collecting digitally created content and tagging as a form of metadata enrichment of the collections, has changed the relationship between museum and its audiences. Digital age has given the opportunity for the museum collections to be more open and visible than ever before. Also experimenting with exhibition formats and enabling visitors to participate, especially in the form of adding their ideas or statements in the form of comments, has become common practice. Although often these activities are

3 The new museology in Peter Vergo's intervention (1989).

not sufficiently modelled or systematically developed, it can be claimed that audience participation has shaped the power relations between museums and their audiences and facilitated partnership. Eventually, this can be expected to give a stronger role of audiences in the decision-making process of the museum.

Media scholar Sonja Livingstone stresses that participation is always mediated and proposes to ask, what modes of participation are afforded to people by the particular media and communication infrastructure which mediate social, cultural or political sphere of life (Livingstone, 2013: 28). In the same vein, it is important to look at how museums can engage audiences, how they understand people's motivation and how people's everyday life and the knowledge offered by the museum are related. The success of participatory actions in the society is based on "when the political becomes personal" (McAfee, 2000: 159–160) or when people's subject positions are intertwined. Cultural heritage, memory or past is not necessarily part of one's everyday life, therefore it is hard to include people into creating heritage and dialogue related to the heritage. Getting people to participate needs empathy and skills to be relevant in people's lives from the museum institutions.

How much and in which ways participatory possibilities have impacted the museum tradition and their view as gatekeepers of knowledge, is hard to show, as participation is a continuous process. To analyse and understand the changing dynamics between museum and its audiences, our research team selected the Estonian National Museum (ENM) as a case study for the research. What makes this institution interesting for study, is this current transformation, outlining its contradictory role at the society. The ENM was created as part of national movements in 1909, established as important symbol of national memory and identity in Estonia. Runnel et al. (same publication, 19–34) have followed the challenges related to the ENMs contemporary identity and explored the values the museum has to offer to modern society at the time when museum has entered a major museum-building project. The research has showed that the museum has many roles and is currently facing contradictory presumptions, based on two quite separate discourses – one popular and traditional, possessed by the public, which sees the museum and its collections as important source of national culture. The second intellectual and academic, is in the possession of professionals. It favours negotiations and sees museum as communication institution (ibid.), strongly challenging the first, popular discourse. Thus it is a challenge to make audience participation part of museum's practices of transformation within this framework.

1. Audience participation in museum as a research method

The aim of his article is to use the case study of the Estonian National Museum for presenting a multi-method and multi-site interventionist approach to initiate and study audience participation in museums. The article is based on a five-year research project investigating museum participation, where instead of only just observing the status quo and comparing that to the theoretical possibilities on participation, we undertook initiatives to change the participatory conditions in the ENM through research. We also set out to overcome the ambivalence offered by two strong discourses about the Estonian National Museum and designed the project to address people's everyday life, thus inviting them to participate as experts of their own lives.

More widely, the project set out to investigate museum communication in the twenty-first century information environment,[4] but also to initiate some communicative action to support the participatory potential of museum communication. We decided to take a holistic and ethnographic approach which facilitated understanding audiences in their actual social contexts. Equally important, the holistic approach demands an understanding of the museum as a non-linear communicator in a participatory situation. In many cases, museums find it difficult to venture beyond very traditional communication situations with their audiences.

As previously discussed, museums in recent times have increasingly been faced with social and communicative challenges. In order to both meet and study these challenges, a research group consisting of people from the University of Tartu and from the ENM was set up. Most of the members were simultaneously connected to both organisations as researchers and PhD students, adding additional challenges in terms of finding insider/outsider balance for the participatory interventions.

This combination of insider and outsider interpretive perspectives allowed the research group to apply both analytical outsiders' view to museum practices and to understand insider's culture and contexts. The methodological decision to choose an insider action research approach was also grounded in the idea that at the end of the research-interventions the staff at the museum should ideally be able to understand different aspects of museum participation.

Our research group was particularly interested in considering an interventionist research project in which some of the proposed or considered changes are at the same time investigated through the research project and enacted as interventions in real-life situations. We will thus continue to discuss some of the founding methodological principles of this project – namely the ideas borrowed from

4 More information about the project can be found at: http://muuseum.edicypages.com.

(insider) action research and the introduction of real-life experiments or interventions to investigate and change museum participation. After that, we will look at the multi-method and multi-site components of these intervention projects, and at the related notions of ethnographic research, data triangulation and team-based reflection.

2. (Insider) action research and ethnographic research

The methodological conceptualisation consists of three layers, each adding an analytical dimension to both data collection and interpretation process. The first layer is connected to the concept of action research or conducting interventions, the second layer brings the focus to the inside of the organisation. The third layer adding mostly interpretational and analytical depth is ethnographic research.

2.1. Action research

On the first layer, the ENM research project can be considered an action research project. Action research integrates the idea of exploratory action into social research. Its purpose is to change possible challenges in organisation through the research. It is an emergent and collaborative, research approach, as it involves working with the people whom you study and aim at improving the system within which your participants work. Action research as insider research means that the researcher works *with* practitioners rather than for them, with the aim of effecting change rather that just studying it (Bradbury Huang, 2010). By Holian the outcomes of action research are both practical and theoretical: the knowledge the research generates has direct and ongoing impact on participants and also wider audiences (Holian et al., 2013). In the area of culture, theatre activist Augusto Boal demonstrated that audience interventions on stage can construct new ways of acting in the everyday life (Osterlind, 2008), which has been also part of the agenda in the research interventions discussed in this article.

The aim of the research project was strongly connected to generating actions that would change the nature of participatory relationship between the museum and its audiences. Researchers of university were part of planning, designing, working and analysing the participatory interventions with the ENM research team, which in our case was only a small group as compared to the whole organisation. True to action research ideas, a series of seminars were run and many aspects of the interventions were debated rigorously with various teams from the museum in order to enable the organisation to get familiar with the participatory intervention as well as to evaluate their course.

Different online and onsite interventions were created by the research team where the principle was to investigate different aspects of participation, for example to study audience content creation in the different conditions of the exhibition space (in forms of comments or story telling), to study how museum responds to sharing knowledge on collections or audience content contributions to the exhibitions. It is always important to ask questions about ethics, when designing action research projects and interventions: who is to be involved; how and why; who makes decisions and how; whose interpretations are to prevail and why; how it will be written about the people involved in processes and who owns the ideas developed (Eikeland 2006). It is important to develop the interventions in cooperation and in partners respect.

2.2. Insider action research

So from the different branches of action research, the second layer to our approach focuses on the insider action research and means that the research and actions are initiated, carried out and analysed by members of the organisation (Coghlan, Brannick, 2001). Insider action research is centred on the process where a full member of an organisation is researcher, who takes in addition to the normal functional roles they hold in an organisation action research role (Holian et al., 2013).

Insider action research does come from a management perspective, mostly from business, health and education with the clear aim to transform the organisation and to study the processes, becoming engaged in the study of one's colleagues, clients, patients or students. Projects usually focus on issues that have been identified and selected by the researcher in collaboration with organisation which are seen as opportunities or problems that need to be addressed and which have a clear possibility to influence the organisation. The researchers and all other organisation members are not merely passive recipients of the intervention, but play an important role in determining the intervention process (Nielsen, 2013).

In our case we could see that there were differences in professional competence, cultural and personal backgrounds, presuppositions and prejudices and personal ways of doing things within the research team. For the research group, besides being a process to follow or an act that interfered habitual ways of doing things or being a change that requires developing new skills, the research interventions were also theoretical and analytical constructions.

In combining the insider-initiated interventions with the outsider influences of university researchers, we included the elements of participatory action research. Participatory action research stresses participation even more than the original thoughts of action research (McTaggart, 1997; Brydon-Miller, Maguire, McIntyre, 2004, quoted in Sundin, 2010).

Specifically, our approach to the insider action research is characterised with the three key concepts of 'pre-understanding', 'role duality' and 'organisational politics', and the possibility that the researcher is also an employee of the organisation where the research takes place (Coghlan, 2001, quoted in Sundin, 2010). We considered it appropriate and relevant to investigate participatory interventions through this type of real-life experimentation.

As such, the insider approach, where the research team is at the same time running the experiments and investigating the results and implications for the museum, provides both advantages and disadvantages. Insider researchers have good understanding of organisation, but they also hold both multiple organisational roles as well as the role of a researcher at the same time. In terms of participation, the researcher is relatively free and can shift from the position of participant to observer and vice versa. However, this shift of position can also prove to be an obstacle to role-balance when the staff members are caught in "loyalty tugs, behavioural claims and identification dilemmas" (Coghlan, 2003, quoted in Dover, 2008). The researchers are visible in organisation, but at the same time they should avoid the role of being "the-one-who-knows".

Researchers might also run into an organisation's "undiscussables" as well as become the target of accusations of spying and self-promotion (Coghlan, 2003, quoted in Dover, 2008). Coghlan argues that these difficulties are more likely to arise in the more "organistic" action research process, which values a process of inquiry that also addresses "underlying assumptions and defensive routines" (Coghlan, 2003, quoted in Dover, 2008).

Therefore, we decided to bring the academic output back to the museum by discussing the research, which allowed the museum professionals to challenge and also to learn from it. This is also where the series of research interventions run by our research group differs from simpler, more pragmatic, more "mechanistic" participatory action research projects. Our project not only focused on the pragmatic outcomes of clear benefit to the organisation, but also on "enacting a transformation of being" (Coghlan, 2003, quoted in Dover, 2008), which is related to the agenda of social and communicative museology.

While insider action research is not something radically new, it is not very often applied in the context of audience studies. The benefit of this interventions initiating approach, however, is in forcing the course of the institution's relations to its audiences in order to study these relations. The method is also very resource consuming from the organisation side as it assumes not only willingness to open itself for research, but to take audience participation seriously to be willing to change with the course of the project.

3. Intervention as a method to investigate participation

As a third methodological layer of the research project, the concepts found within ethnographic research share common ground with those found within insider action research. From the principles of ethnographic studies, our research team saw the interventions as one way of creating situations that would increase the museum's reflexivity about participatory communicative situations, as well as reflexivity over the research agenda and research processes – for example the issues of ethics, use of technologies and role of the researchers. Moreover, members of the research team have participated in these action-led processes in double roles: as museum professionals and, from a certain point onward, as ethnographers conducting participatory observation "at home", looking closely at the impact of the interventions and taking advantage of being immersed in the culture.

The project was introduced and discussed within a broader group of ENM professionals (open museum board meeting, research department internal seminars), as well as at the actual implementations, during related exhibitions, and in Web-based interactions – all this in order to carry out the principles of the insider action research and foster knowledge dissemination within the organisation. On top of that, roundtable debriefings were held among the involved and interested museum staff after the first data collection pilot for the storytelling intervention "Give the Museum A Day" in 2009 and when the interventions dealing with exhibition production through the open curatorship project "Create Your Own Exhibition" had finished.

These interventions could also be considered one-shot case studies (Campbell, Stanley, 1973) or natural experiments (Babbie, 2010). This means that the interventions took place at the museum in the real-life situations, in actual everyday work environment with its complex of practices and relations, with possibility to follow how the interventions influenced actual relations with the audience and co-workers, how it affected the objects and how the interventions shaped the scenes of everyday life. The key concept of the experiment – i.e. providing a stimulus and exploring the consequences – remains the same as in laboratory conditions, however the effects are more difficult to evaluate in the natural experiment conditions as it is harder to understand the causality of the event. The only possible comparison afforded by this kind of research situation is through comparing a set of related and to some extent similar repeated cases. The aim of repetitions and modifications is to provide possibilities to examine the different affordances in each experiment situation. In the context of our research, a total of seven case studies were conducted with additional spin-offs that also provide to some extent comparable data.

Altogether 8 interventions were carried out in the period of 2009–2012, with a mix of museum key questions such as the targeted museum functions and processes, the different techniques and technologies used, and the variety of audience groups targeted (Table 1). The interventions ranged from storytelling to exhibition proposals, from item recommendations to replicating the actual museum items by handicraft communities. The interventions included activities like active visitors commenting museum objects, stakeholder online community representatives re-interpreting museum objects, or active audiences contributing content about their everyday life (description of one's ordinary day).

In a more structural intervention, the audiences were invited to propose and decide on ideas for a do-it-yourself exhibition. Table 1 provides an overview of five intervention cases. The table only shows a sample of interventions and of the case-study analysis framework employed to investigate the different interventions, yet it enables to get some flavor of the interventions inventory.

Our research interventions do carry a multitude of aims. On the one hand, the interventions were designed to challenge the museum, to inspire museum professionals to notice the social and communicative aspects of their institution. On the other hand, these activities have given a multitude of opportunities to study audiences, their understanding of the roles of museums in society, and their conceptualisations of museum participation.

While on many occasions the research focus was on the inside, i.e. towards understanding the museum professionals' identity processes and changing their perception of participatory and communicative museum (Tatsi, 2013), the aim still was opening the museum to the audiences and understanding them better through participatory processes. The staff members were fully aware of the fact that about a thousand of visitors were "reached" by the different participatory initiatives,[5] providing their input, joining the activities, voting for exhibitions or crafting their own version of heritage items, and that many more have been in contact with the results of these experiments by viewing, reading and interpreting the materials produced by the participants.

4. Multi-method approach in data collection

Following the principles of ethnographic research, we considered it very important to have multiple points of data collection – from the meetings setting up and designing the interventions, from the meetings discussing the interventions among the museum staff and from the actual interventions themselves. The par-

5 In addition to that, more than 3300 schoolchildren took part of a drawing competition, which was part of the participatory initiatives held at the "Shopping Fever" exhibition.

ticipatory interventions conducted at ENM were different in nature – in terms of the museum functions they addressed, their reach and their influence on the participants and museum staff. To fulfill the aim of the project for each intervention, the research team has mapped the interaction design, implementation, process and outcomes. The team has also estimated the impact of the action on the museum and on participants. In order to evaluate the different elements of collected data, each category was analysed as seen in the examples provided in Table 1.

Also multiple data collection methods were used to collect feedback from participants, depending on the character and features of the participatory intervention. In the case of online participation, we used online questionnaires, e-mail interviews and public polling. The researchers also investigated public for in order to understand specific target groups' opinions about the museum and its activities. For on-site participation, paper questionnaires, storytelling, paper-based polling, interviews with participants and observation were used. The aim of the multi-method approach was twofold. On the one hand, we were interested in collecting data on the participants' experience with, and expectations of museums. On the other hand, the data collection methods were aimed at collecting the participants' impressions about the participation processes, and their motivations to participate. At the same time, the participants' social data and background were collected in order to map the different profiles of participants.

The diverse and to a certain extent ad hoc nature of the project meant that for each intervention the choices of which data and how to collect data were made together with the intervention design. If the particular action needed greater contributions from participants (as in the case of "My Favorite Item in ENM's Collections"), interviews with participants were used in the end of the action, whereas when participation was a side effect of a visit and framed as interaction in the economic field (as in the case of exhibition commenting projects), observations or surveys were used. In all cases of working with the intervention data, both participants' contributions and the reflections from the museum professionals were analysed.

There is a lack of consensus of how participation in intervention situations should be measured. In management or health studies some use qualitative approach (Nielsen et al., 2006; Aust et al., 2010), others have attempted to develop quantitative measures either measuring the quantity or quality of participation (Nielsen, 2013). Simon (2010: 301) has stated that a lack of good evaluation of participatory projects is probably the greatest contributing factor to their slow acceptance and use in the museum field. Constant evaluation of the participatory interventions at ENM was set as one important goal of the research project.

Table 1: Section of the interventions inventory

Data framework	Key questions	***Give Museum a Day from Your Life*** **(April 2009)**	**Exhibition Commenting with Pen and Paper (fall/winter 2010)**	**Open Curatorship Exhibition (2010–2011)**	***My Favorite Item in ENM's Collections*** **(winter 2011)**	***Shopping fever*** **(2012)**
Description	Overall aim	Collect stories of everyday life of Estonians on ENM 100th Anniversary	Receive comments and clarifications on photo collection displayed on exhibition	Hosted exhibitions invitation to general public to display their own collections or cooperation with museum collections	Invite handicraft community to remake museum items either as authentic copies or inspiration items	Develop series of participatory events covering different aspects of the exhibition on the emergence of consumer society
	Museum process/ function	Collecting	Displaying (and collecting)	Displaying/partnership/ collecting	Remix/collecting/displaying	Remix/collecting/displaying
	Participant activity	Telling stories	Commenting, giving opinion	Exhibition proposals, voting, holding exhibitions, giving collections to museum.	Remake museum items, telling stories	Suggesting items for collections, voting, submitting drawings and posters (schoolchildren) telling their experiences of disappointing purchases, interacting with other visitors, co-designing part of the exhibition
	Access point: Online	Advertisement online, e-mail address and online form as submission sites	Advertisement of exhibition, no participation possible	Online submission of exhibition proposals. Online evaluation of exhibition proposals by public vote	Advertisement in different forums. Online submission of interest in participation and online submission of finished works. Use of online collections for inspiration	Online contribution of photos, discussing with other contributors and voting for favourite photos

Data framework	**Key questions**	***Give Museum a Day from Your Life*** **(April 2009)**	**Exhibition Commenting with Pen and Paper (fall/winter 2010)**	**Open Curatorship Exhibition (2010–2011)**	***My Favorite Item in ENM's Collections*** **(winter 2011)**	***Shopping fever*** **(2012)**
Description	Access point: On site	Postal and hand-delivered stories	Post-it notes stuck to favorite pictures worth commenting	Offline submission of exhibition proposals. Offline display and evaluation of proposals. Two winning exhibitions displayed at the main museum building	On-site submission of finished works	On-side submission of objects, stories, drawings, posters, co-designing part of the exhibition by a small group of visitors
Statistics (example)	No of comments: Online	23 online submissions	NA	28 online proposals, 509 online votes	47 people registered, 37 people completed work. Viewers of online exhibition	319 online contributions of photos (711 submitted photos); 50 stories on regretted purchases, 44 submitted objects; more than 3300 drawings by schoolchildren
	No of comments: On site	402 on site or postal submissions	80	5 on-site proposals, 55 on-site votes	~ 10 people brought in their work and only registered later online. Viewers of selected items in museum	12 016 views of related comments at the campaign's Facebook page
Impressions (example)	Challenges	How to give feedback to participants that creates sense of relevance? Do these kinds of collection activities have to have concrete output – exhibition, publication?	How to sustain the comments and foster interaction? How to invite people to revisit their comments and create dialogue?	Balancing different interests, "populist" voting, how to realise the potential of engaging new types of audiences?	What impact would have parallel public voting had to the process? The access to the actual collections and items within the collections can be a bottle neck, as people need guidance and only a limited number of people can actually access the collections at a time	How to distinguish online and offsite as channels, tools and entry points, related to the specific social contexts of the audiences; how to design and handle online participatory event through specific web page, linked also to social media

Data framework	Key questions	*Give Museum a Day from Your Life* (April 2009)	Exhibition Commenting with Pen and Paper (fall/winter 2010)	Open Curatorship Exhibition (2010–2011)	*My Favorite Item in ENM's Collections* (winter 2011)	*Shopping fever* (2012)
Analytical (example)	Access	Increased sense of relevance – your story matters, you should be "preserved" at the museum	Visitor has potential access to the captions of the photos in the collections, can edit and suggest changes. However, it was not made explicit whether and how the collections would use that data	Access to museum space is affected	Collections are open-access and increasingly available online. The access to the museum "prestige" and expertise of national culture and to peer-feedback. Access to discussions	Visitors are able to access and use the exhibition as the platform to tell each other their stories and experiences related to the emerging consumer society from their everyday life perspective
	Interaction	Interaction with new groups, new communities who saw the museum's relevance	Visitors could interact with the curated content as well as with each other's capitation	Interaction between audiences, participants, research team and museum workers	People participate gladly in interpretations of national heritage and are prepared to work for the museum in copying. Innovative engagement in openness of the interpretation process	Visitors could interact with collections, exhibition content and with each other regarding their own contributions
	Participation	Minimalist contributive participation addressed and accessioned to the museum collections	Minimalist contributive participation in exhibition space, no permanent impact on collections	Maximalist hosted participation in exhibition space. Limited accession impact on collections	Minimalist contributive participation. Innovative accession impact on collections	Mostly minimalist contributive participation, but also a maximalist participatory action for school children for co-designing part of the exhibition itself

In order to systematise the data collected for each intervention, store the information and analyse it later a single framework was used. The framework was composed of four sections: (1) a description of the intervention, (2) statistical information, (3) the organisers' and participants' impressions of the project and (4) an impact analysis (Table 1).

The description of each intervention included the following items: (1) the different goals of the intervention project, distinguishing between research goals, participation goals and museum goals, (2) the target group and the promotion plan, (3) the possibilities of access (online and on-site) to the intervention, and (4) the description of the intervention process itself. The description has provided important background information for later analysis in order to determine possible success or failure factors. Data was based on project leader materials and on museum staff meetings. The aim here was to store as much of the intervention related data as possible for the future repetitions, but also to be able to distinguish the elements leading to success or failure.

The statistical section of the framework included information on (1) the costs of the intervention, (2) the project duration, (3) the number of participants, (4) the preparation time and staff, (5) the proposed incentives to participate, and (6) the outcomes of the participatory intervention in the form of actual products (e.g. the number of stories collected, the number of items replicated). Statistical information has provided factual data that enabled cost-benefit analyses – i.e. it has offered possibilities for the interested parties to estimate whether the cost/effort and outcome balance might be desirable. Collected data was based on project leader materials. While this data could also be seen as part of the design and implementation question, then the first section was more impression based and later tried to gather all statistical information possible.

When it came to analysing the impressions of the project, we focused on (1) usability, in the sense of ease and comfort of use of the participation facilities (Nielsen, 2006), (2) the participants' behavior, (3) the participation process and the evaluation of how the participants cooperated, (4) the successes and failures of the participation management, and (5) the benefits of the intervention for the museum and for the participants. Especially important here were the constraints, focuses and obstacles imposed on participants, as well as the practical failures of the participatory process. The impression section has supported team reflexivity and internal communication. Collected data was based mostly on project leader materials, museum staff meeting materials, interviews or other collected feedback from participants. Here the multiple data allowed gathering diverse viewpoints.

The impact analysis of the intervention addressed several questions: (1) Who was empowered or limited in terms of access, interaction and participation (Carpentier, 2011); (2) Who was affected – the participants, the museum professionals or the intervention facilitators?; and (3) What was affected – the interactions between the participants, the objects (collections, exhibitions), the processes (i.e. the working practices), or the museum institution? Thus the analysis has attempted to pinpoint the affected aspects of the museum. For instance, the collections were affected when the outputs of the participatory interventions were actually included in the museum's collections, and the working processes were affected when such inclusion required re-thinking of the established collection practices.

5. Online and on-site

As the participation process is facilitated by the digital networked media, different interventions added online and on-site components to the investigation. The online space was in many ways used to broaden the museum audiences – to include groups otherwise left out of museum activities. Also when the intervention was designed for audiences using web daily, we considered important to also keep the possibilities for offline participation to give everyone access to the participation, as in the case of the campaign "My Favorite Item in ENM's Collections".

The interventions were advertised through online channels with the visibility of these activities being high in the museum space as well. None of the activities conducted within the framework of this project remained only online. The aim was to engage the different spaces as extensions of each other, providing information and incentive as well as the possibility of online participation, in addition functioning as an invitation to the on-site museum space. Thus, the online space combined this invitation with inclusion in museum activities. For example, in the "Curate Your Own Exhibition" intervention, people were invited to submit their proposals both online and on-site, and the possibility of voting for the exhibition was also provided both online and on-site. Both submission and voting processes also worked as incentives to come and visit the actual museum site when the exhibition was held in the temporary exhibition room.

Photo competition "Take a Photo of what You are Eating" took place online and participation – submitting and commenting on photos. The photo exhibition was also shown in the museums as a side project to "Shopping Fever" exhibition in order to give participants feeling that they are not just part of online, but also participate at the real exhibition in the museum. Mostly the participation still preceded the site visit. Similarly experienced participation the participants of the drawing exhibition "My Gift", where sending in the drawings was followed by the on-site exhibition, where all the submitted drawings were displayed.

Although Jakob Nielsen's (2006) framework for understanding and analysing the usability of websites was initially designed to support people attempting to build online communities, it also supports the analysis of participatory initiatives, as practitioners can then look at the strengths and weaknesses of their activities. While Nielsen focuses mainly on issues of online engagement and usability in general, the ideas proposed, and the support provided, also suit on-site activity. Addressing issues of usability supports participation through simplifying the process for the participants. Nielsen argues that there are five key possibilities for increasing user participation in content creation: (1) making participation easier, (2) providing the possibility to edit rather than create, (3) promoting quality contributors, (4) making participation a side-effect of visiting the site, and (5) rewarding, but not over-rewarding, participants.

The participatory interventions at ENM have not always followed these recommendations, although we have managed to implement many of them, especially using online initiatives. However, in hindsight there were measures that could have been implemented better. The quality of the contributions is one of these usability aspects that could have been promoted to greater effect, when providing examples or editing possibilities. Another lesson learned from contrasting the different interventions was that in some of the cases the advertising to the right audience of the participation projects could have been more effective – supposedly that would have increased the number of participants.

The ways museums use technical measures to support or ease participation have to be linked with what the museum has set as an aim for that particular participatory action. When participant numbers are sought, easing participation is very important, although at the same time this ease can become a barrier to more complex or diverse contributions. Sometimes audiences see the barriers as adding value to participatory initiatives, while at other times an expert jury or real-life exhibition may become hindrances to participation. Our multi-method data analysis of different experiments has highlighted both such occasions.

No matter if the participation is high-tech or low-tech, technology should not become a barrier in itself. In today's world, this means replicating participatory initiatives online and on-site because some technologies are more accessible to some groups, and other technologies to other groups. In our experience, intertwining works best when engaging a diversity of groups.

6. Conclusion

Overall, although museum audiences are hard to capture, experiments that engage audiences in participatory activities within the museum space provide an important way in which to research audiences.

The use of insider action research in museums thus enables to understand the multiple faces and possibilities of participation, and also to understand the museum audiences in participatory situations. Moreover, the experiment situation, initiated in the museum through theoretical analysis and focused on the interaction between the organisation and its audiences, provides grounds for participatory communication, enabling innovative approaches to audiences and bringing them closer to the museum.

The rise of social media has brought forward public expectations of increased dialogue. Online media supposedly bring audiences and institutions closer to each other by providing more (more or less controlled) opportunities and spaces for interaction. However, as the ENM project demonstrates, these expectations are also applicable to on-site communicative situations, meaning that people are not only content to participate or contribute online, but they also have aspirations to impact on the museum's space and collections.

Our case studies were based on multiple data collection and aimed to summarise the relevant information gathered during each particular intervention. The analysis for each of the interventions, based on the single analytical framework outlined above, enabled the comparison of otherwise fairly different actions. The same framework was used by other organisations for analysing their participatory activities, which has given many comparison opportunities beyond the ENM project.

The interdisciplinary background of the researchers involved in this study has enabled the merger of different methodological approaches and conceptualisations of the audience. The strength of this project is that it brings together participants with diverse interests located in the different fields discussed above. Thus, a participatory intervention can enable the targeting of specific audience groups who are willing to become engaged and support the museum in its activities. The interventions have provided the museum with feedback opportunities and possibilities to engage in dialogue. The different angles from which the museum and its audiences were interrogated provided a multitude of insights. Methodologically, the novelty of this approach is mainly down to using an insider interventionist approach in order to initiate and investigate transformations of both the audience and the institution.

We can suggest that when organisations or researchers develop interventions or action research they need to consider that their planned activities can be realised with the available resources and in the context of the particular museum organisation. This needs new skills and openness – the researcher has to be able to act also as a manager, producer and communications officer.

The methodology of insider action research does not enable to measure how strong was the audience involvement with cultural heritage or how strongly people found they are involved with creating new knowledge in the museum settings when participating in the interventions. However, the research interventions challenged the traditional roles of producers and consumers, experts and non-experts at the ENM also outside the particular interventions, by foregrounding cooperation and showing the importance of participation as possibility to communicate with audience within the organisation.

Acknowledgements

The publishing of this article was supported by the Estonian Science Foundation grant no 8006.

References

Aust, Birgit; Reiner Rugulies, Annett Finken, Chris Jensen (2010). When Workplace Interventions Lead to Negative Effects. Learning from Failures. *Scandinavian Journal of Public Health* 38 (3): 106–119. Quoted in Karina Nielsen (2013). How Can We Make Organizational Interventions Work? Employees and Line Managers as Actively Crafting Interventions. *Human Relations,* 66: 1029–1050. Available at: http://hum.sagepub.com/content/66/8/1029 (accessed September 30, 2013).

Babbie, Earl R. (2010). *The Practice of Social Research.* Wadsworth: Cengage Learning.

Barrett, Jennifer (2011). *Museums and the Public Sphere.* Chichester: Wiley-Blackwell Publishing.

Bradbury Huang, Hilary (2010). What is Good Action Research?: Why the Resurgent Interest? *Action Research* 8 (1): 93–109.

Brydon-Miller, Mary; Patricia Maguire, Alice McIntyre (2004). *Traveling Companions: Feminism, Teaching and Action Research.* Westport, CT: Grenwood Publishing Group. – Quoted in Ebba Sundin (2010). Action Research in Media Studies. In Nico Carpentier et al. (eds.). *Media and Communication Studies Intersections and Interventions*. Tartu: Tartu University Press, pp. 267–276.

Campbell, Donald T.; Julian C. Stanley (1973). *Experimental and Quasi-Experimental Designs for Research.* Boston: Houghton Mifflin Company.

Carpentier, Nico (2011). *Media and Participation. A Site of Ideological-Democratic Struggle.* Bristol, UK: Intellect.

Coghlan, David (2001). Insider Action Research Projects: Implications for Practising Managers. *Management Learning,* 32 (1): 49–60. Quoted in Ebba Sundin (2010). Action Research in Media Studies. In Nico Carpentier et al. (eds.). *Media and Communication Studies Intersections and Interventions*. Tartu: Tartu University Press, pp. 267–276.

Coghlan, David (2003). Practitioner Research for Organizational Knowledge: Mechanistic- and Organistic-Oriented Approaches to Insider Action Research. *Management Learning,* 34 (4): 451–464. Quoted in Graham Dover (2008). The Importance of Working on the

Inside. *Participation Action Research & Organizational Change*, July 10. Available at: http://participaction.wordpress.com/2008/07/10/insider/ (accessed August 29, 2012).

Coghlan, David; Teresa Brannick (2001). *Doing Research in Your Own Organization*. London: Sage.

Dana, John Cotton (2004). The Gloom of the Museum (1917). In Gail Anderson (ed.). *Reinventing the Museum. Historical and Contemporary Perspectives on the Paradigm Shift*. Lanham, New York, Toronto, Oxford: AltaMira Press, pp. 13–29.

Eikeland, Olav (2006). Condescending Ethics and Action Research: Extended Review Article. *Action Research,* 4 (1): 37–47. Available at: http://arj.sagepub.com/content/4/1/37.

Holian, Rosalie; David Coglan (2013). Ethical Issues and Role Duality in Insider Action Research: Challenges for Action Research Degree Programmes. *Systemic Practice and Action Research,* 26: 399–415.

Hooper-Greenhill, Eilean (2011). Studying Visitors. In Sharon MacDonald (ed.). *A Companion to Museum Studies*. Oxford: Blackwell Publishing Ltd.

Jenkins, Henry (2006). Confronting the Challenges of Participatory Culture: Media Education for the 21st Century (Part One). *Confessions of an Aca-Fan: The Official Weblog of Henry Jenkins*, October 20. Available at: http://henryjenkins.org/2006/10/confronting_the_challenges_of.html (accessed August 29, 2012).

Livingstone, Sonia (2013). The Participation Paradigm in Audience Research. *The Communication Review,* 16 (1–2): 21–30.

McAfee, Noelle (2000). *Habermas, Kristeva, and Citizenship*. Ithaca: Cornell University Press.

McTaggart, Robin (ed.) (1997). *Participatory Action Research: International Contexts and Consequences.* Albany: State University of New York Press.

Nielsen, Jakob (2006). Participation Inequality: Encouraging More Users to Contribute. *Jakob Nielsen's Alertbox*. Available at: http://www.useit.com/alertbox/participation_inequality.html (accessed September 30, 2013).

Nielsen, Karina; Hanne Fredslund, Karl B. Christensen, Karen Albertsen (2006). Success or Failure? Interpreting and Understanding the Impact of Interventions in Four Similar Worksites. *Work & Stress,* 20 (3): 272–287. Quoted in Karina Nielsen (2013). How Can We Make Organizational Interventions Work? Employees and Line Managers as Actively Crafting Interventions. *Human Relations,* 66: 1029–1050. Available at: http://hum.sagepub.com/content/66/8/1029 (accessed September 30, 2013).

Nielsen, Karina (2013). How Can We Make Organizational Interventions Work? Employees and Line Managers as Actively Crafting Interventions. *Human Relations,* 66: 1029–1050. Available at: http://hum.sagepub.com/content/66/8/1029 (accessed September 30, 2013).

Osterlind, Eva (2008). Acting out of Habits – can Theatre of the Oppressed Promote Change? Boal's Theatre Methods in Relation to Bourdieu's Concept of Habitus. *Research in Drama Education,* 13 (1): 71–82.

Runnel, Pille; Taavi Tatsi, Pille Pruulmann-Vengerfeldt (2010). Who Authors the Nation? The Debate Surrounding the Building of the New Estonian National Museum. In Simon J. Knell et al. (eds.). *National Museums. New Studies from around the World.* London; New York: Routledge, pp. 325–338.

Simon, Nina (2010). *The Participatory Museum.* Santa Cruz, California: Museum 2.0.

Stylianou-Lambert, Theopisti (2010). Re-conceptualizing Museum Audiences: Power, Activity, Responsibility. *Visitor Studies,* 13 (2): 130–144.

Tatsi, Taavi (2013). *Transformations of Museum-embedded Cultural Expertise.* Dissertationes de mediis et communicationibus Universitatis Tartuensis 18. Tartu: Tartu University Press.

MUSEUM AUDIENCES AS PARTICIPANTS

Handicraft Hobbyists in an Ethnographic Museum – Negotiating Expertise and Participation

Krista Lepik

Pille Pruulmann-Vengerfeldt

1. Introduction

Relatively recently, Desvallées and Mairesse stated that "the chronic lack of interactivity in museum communication has led us to ask ourselves how we can make the visitor more active, while seeking his participation" (2010: 30). This article, which looks at visitor participation, focuses on a very specific group of visitors, handicraft hobbyists, and more specifically their relationship to an ethnographic museum, the Estonian National Museum, by asking members of this specific group what museums do and should be doing in order to make use of visitor input.

This chapter makes its contribution by focusing on museum-goers' perception of participatory practice. It departs from the constructivist, grounded theory developed by Charmaz (2006), and pays attention to concepts that are important to the visitors, and to their view of the role of the museum in their lives. Against the backdrop of earlier works (Goodnow, 2010; Simon, 2010; Pruulmann-Vengerfeldt, Runnel, 2011), this paper attempts to provide a direct answer to the question: What do the visitors themselves think about cultural participation in museums?

The works mentioned in the previous paragraph have contributed extensively to our understanding of various forms of participation in museums, and how these forms might be compared to one another. For example, Goodnow (2010) takes a somewhat hierarchical approach, relying on Carpentier (2007), and delineates participation at the levels of access, reflection, provision and structural involvement, on the basis of the extent of power handed over to participants in museums. Instead of treating various forms of participation "as progressive steps" (Simon, 2010: 188), Simon (2010) suggests considering different variables that help to distinguish "contribution, collaboration, co-creation, hosted" (ibid.: 188) models of the participatory museum. Different variables (e.g. power handed to potential participants, the institutional commitment, the motivation of participants, resources, skills and eventually the perception of non-participating visitors (Simon, 2010)) all come together to form a matrix that helps to explain the nature of participatory projects. It is, however, crucial to keep in mind that not all the characteristics of these variables match perfectly with any particular form of real participation. On closer analysis, it emerges that different projects borrow elements from various forms of participation. The third approach mentioned above (Pruulmann-Vengerfeldt, Runnel, 2011) draws on various fields (economic, political, and cultural) in order to provide an analysis of the possibilities of participation frames within "the classical communication model of Who? Says What? To Whom?" (Lasswell, 1948; McQuail, Windahl, 1993 cited in Pruulmann-Vengerfeldt, Runnel, 2011: 16). With a measure of caution, this latter work also introduces hierarchical models of participation assembled from different disciplines, but its main contribution is that Pruulmann-Vengerfeldt and Runnel (2011) outline a great range of participatory practices in museums. These fields (economic, political, and cultural) are interwoven in practice, and therefore analysing participatory activities may in some cases be more understandable, to both practitioners and potential participants, in terms of categories that are more closely related to everyday practice. As this chapter focuses on the relationship between museums and potential participants, it is also important to theoretically outline the notion of expertise and its relationship with participation.

2. The role of experts in cultural participation

This section looks at expertise in general, then at the area of cultural participation, and finally considers the role of expertise in museums. On a very general level, we can draw on the work of Anthony Giddens (1991). Giddens has emphasised the role of expert systems in contemporary, reflexive society. According to him, "expert systems bracket time and space through deploying modes of technical knowledge which have validity independent of the practitioners and clients who make use of them" (Giddens, 1991: 18). Expert systems (and we see museums as expert systems, too) in this approach are not so much about the power they involve, but rather their scope and knowledge. Indeed, as Giddens himself mentioned earlier, the involvement of communication, power and sanctions is fundamental to all social practice (Giddens, 1979: 82). It allows us to treat the existence of power as a default characteristic, and thus also to leave it in the background, so that we can instead pay attention to other immanent traits that define expertise. For the purposes of this work, it is important to mention that "even the most cherished beliefs underlining expert systems are open to revision" (Giddens, 1991: 141), thus even being "routinely available to laypeople as part of the reflexivity of modernity" (ibid.). These democratising tendencies, after spreading from "the orthodox political arena" (Giddens, 1994: 192) to other domains, have to some extent also influenced the cultural sphere.

In some cases (such as when examining the phenomenon for statistical purposes (Morrone, 2006)), cultural participation is seen as cultural consumption, rather than as something that revises "beliefs underlining expert systems" (Giddens, 1991: 141), or refers to amateur production – "professional practices are excluded here" (Morrone, 2006: 7). In these cases, "cultural participants" are clearly distinguished from "experts", both semantically and practically. These denominations, while designed for statistics and mainly for closed circles of decision-makers, are also made public through the news media and various reports. Thus, they also carry the potential to shape public awareness about cultural participation and shape the opinions of potential participants.

In specific cultural institutions, including museums, a somewhat different picture develops, as the expertise is provided by these institutions but also opened up, to a certain extent, for revisions. Such cultural transformations, however, take time, as both museums and their visitors are deemed to be seeking forms of participation that satisfy mutual expectations (and those of spectators, as Simon (2010) has proposed). The identity struggles of museum professionals as experts were highlighted very recently (Tatsi, 2011) in the instance of an Open Curatorship project staged at the Estonian National Museum, which showed once again the importance of acknowledging participatory practice at both the rhetorical and the practical level. In this case, establishing a participatory intervention triggered debates among museum professionals, as it ran counter to their "traditional" understanding of curatorship (in which the expert alone decides upon the content of exhibitions), fuelled anxiety, resistance and othering, resulted in them focusing on how clearly the visitors' "'amateurishness' becomes evident through the exhibition" (Tatsi, 2011: 73). The borderline between museum and visitors is thus clear and strong even in the case of a participatory project that supports the clearly distinguished identities of museum professionals and visitors. This has the potential to foster a "relatively isolated culture of hosted exhibitions" (Tatsi, 2011: 75) in the future. Although it is for each museum to decide what participatory activities are appropriate for it, the question of striving for mutual exchange of expertise still remains.

3. Context and method

Although this study considers the Estonian context, the issue of lack of interactivity, and the recent signs of will to solve it through participatory initiatives, have been much debated in the European and North American museum communities. Thus, the purpose of this context section is to refer to common traits and issues that the Estonian National Museum (and also other Estonian museums) shares with its international counterparts.

The data for this study consist of interviews and analysis of online materials collected during an intervention study conducted in the winter of 2011. The aim of the study was to involve handicraft hobbyists in reproducing cultural heritage materials found at the museum, either as an authentic copy or as an inspired item. Altogether 47 people indicated their interest by registering, and 37 completed works were submitted for the competition. The entries were evaluated in two categories – copies of originals and inspired items. The evaluation was carried out by a jury, consisting of museum staff and experts invited from the local community. For the data collection, nine interviews were conducted during and after the event (Table 1). The respective quotes are marked with a number (I 1–9) to indicate the interview. Additionally, as the hobbyists were invited through their online community forums, material from those forums (a total of 23 forum topics with 370 posts) and related blog posts (nine posts in total) were collected during and after the competition.[1] Those posts were not used for detailed analysis, but as contextual information accessed through close reading.

Table 1: Interview participants and their related competition works

Interview Code	Gender	Age	Competition entry
I1	F	44	Karja quilt (Figure 1)
I2	F	63	Gloves *Luigi*
I3	F	45	Mittens
I4	F	34	Ceramic dessert bowls
I5	F	38	Chamber of pins
I6	M	32	Hammer-wrought tools
I7	F	42	Ram skin pouches
I8	F	33	Bag in Tunis technique
I9	M	54	Wrought rack for herring baking

For the purposes of analysing the interview data, the constructivist grounded theory developed by Charmaz (2006) was applied. This implies that stress was placed on a "participant's definitions of terms, situations, and events" (Charmaz, 2006: 32), while focusing on "his or her assumptions, implicit meanings, and tacit rules" (ibid.). The handicraft hobbyists' understandings of relationships and collaboration with the ethnographic museum, in terms of trust, access, degree of control etc., are mostly embedded in these assumptions and meanings. In order to gain a

1 The authors are grateful to Master's student Marke Teppor, who was responsible for the running of the intervention, related data collection and initial analysis (Teppor, 2011) in the framework of her thesis project.

better insight into the perceptions of the museum held by the handicraft hobbyists, line-by-line coding (Charmaz, 2006) was applied – this made it easier to understand that the identity of handicraft hobbyists as museum-goers can very much be analysed through meanings they attribute to the museum itself. The ways in which the museum is identified in the interviews give "points of departure" (Charmaz, 2006: 17), and evoke certain differences and similarities that help to position the identity of hobbyist crafters as museum-goers. Various differences and similarities, then, are used in the process of axial coding as "conditions, the circumstances or situations that form the structure of the studied phenomena" (ibid.: 61). These conditions influence potential ways of participation in the museum that, in terms of Charmaz, can be seen as the "actions/interactions, participants' routine or strategic responses to issues, events, or problems" (ibid.), and that eventually can lead to certain "consequences, outcomes of actions/interactions" (ibid.) – either material (such as the tangible results of some common project, for example) or mental (the experiences from the participatory process, "feeding" in new conditions supporting or hindering participatory processes in the future).

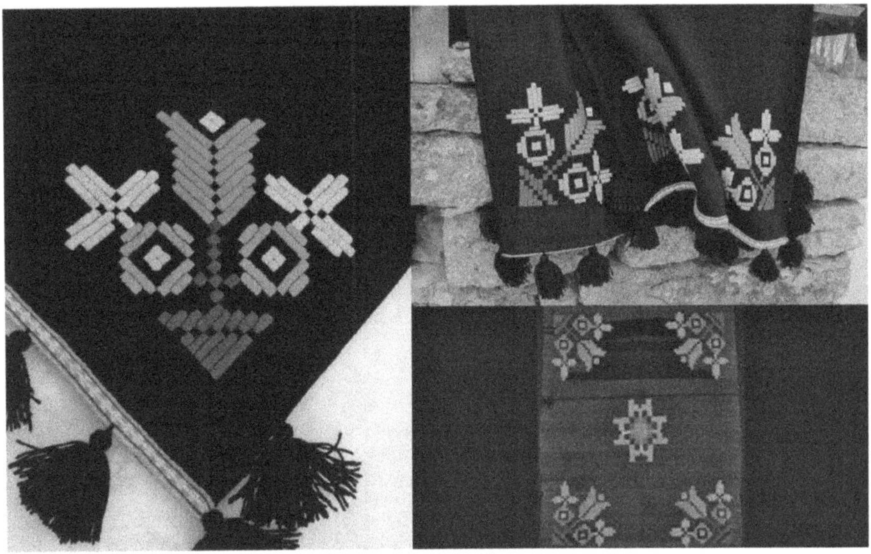

Figure 1: Competition entry in the inspired item category (original in the right-hand corner), author I1 (Teppor, 2011)

4. Estonian hobbyist crafters' perceptions of the Estonian National Museum's expertise

Various roles played by the Estonian National Museum, as perceived by participants of *My Favourite*, have been analysed. It appears that, besides the articulation of the traditional tasks of museums (namely acquiring, preserving, researching, communicating and exhibiting (ICOM, 2006)), a dimension reflecting the expertise of a museum is also present. The expertise seems to be grounded in four distinguishable characteristics displayed by the museum: its large scale, its possession of cultural treasures, its knowledge and its management of risks or conflicts (Table 2).

Table 2: Characteristics of the Estonian National Museum according to handicraft hobbyists

	The large scale	**Cultural treasures**	**Knowledge**	**Conflict or risk management**
Acquiring	Temporal dimension: reaches back to the past; spatial dimension: Estonian and Finno-Ugric culture	The ENM is best aware of what is to be collected, has the right set of values	Knowing what is the "proper" item to be collected	What is to be acquired today? What is authentic and what is fake?
Preserving	Temporal reach to the future, for generations to come	"Depositing" heritage	Knowing how to take good care of old, delicate objects	Conflict between preserving and exhibiting, finding solutions
Research	Large collections nourishing research	Research of treasures	Knowledge taken for granted: "museum knows best..."	Are researchers the only people to be allowed to work with original artefacts?
Communication	Large collections provide plenty of information for exhibitions, educational events and for studying collections individually	Introducing delicate works, popularising old toys (to counterbalance the impact of mass production) and archaic craft techniques	Knowing how to organise an exhibition, distinguishing "good" ideas (about what is to be exhibited) from less good ones	What artefacts should be digitised first (in order to improve public access to information)? Quality *vs.* quantity of digitising

It is, therefore, relatively easy to depict the Estonian National Museum's identity (as perceived by handicraft hobbyists) in a brief table. Since, in some cases, the boundary between communicating and exhibiting can be quite thin, and it is possible to view exhibiting as a part of communication (as in the "PRC model" (Reinwardt Academie cited in Desvallées, Mairesse, 2010: 68)), here exhibiting is also present in the category of communication.

The large scale of the museum is evident both in temporal and spatial dimensions. The temporal reach is considered important, as the ENM is related to objects from "old times", from the past, introducing them to current visitors and future generations. At the same time, the ENM also displays large scale spatially, as handicraft hobbyists refer to the folk traditions inherent to the resources of an entire ethnos, and of all walks of life. The large scale of the museum is beyond the grasp of an individual. Therefore, handicraft hobbyists highly value the information that is made accessible to museum users who need to study the collections individually in the museum's study rooms: "It is very pleasant that these are on display for interested people and craftsmen. So that lay people who do not conduct scientific research there are allowed to come up close and have a look. This is very, very positive." (I4)

The large scale also poses a problem for handicraft hobbyists, as the collections of the ENM are "immeasurable, but there's not much information about the contents of the collections, of what could be found there" (I5). This means that more communication about the scale and richness of the collections is expected, and, despite the scale, a degree of availability is also expected.

Possession of *heritage as a cultural treasure* is the second important aspect of the ENM as an expert. On the one hand, the value of this treasure is hidden in relative all-inclusiveness (as the ENM is interested in Estonian and Finno-Ugric culture), in the quantity of the museum. On the other hand, handicraft hobbyists also emphasise the quality and exquisite essence of cultural treasure. Here, the critique of contemporary mass production or crafts performed slovenly or in a hurry is notable, as is the wish to learn from high-quality items created by previous generations. So we meet the same centuries-old paradox that Gauntlett (2011: 48) has described: "the Arts and Crafts alternative led to beautiful handmade products that the typical worker could not afford", and that can only be eliminated by "doing it yourself". When talking about "cultural treasures", the interviewees usually remain quite generic about particular methods of communicating, yet they emphasise the purpose of introducing "cultural treasures": "popularising old toys for children, to counterbalance" the impact of mass-produced toys (I3), "introducing archaic techniques of work and maintaining a distance from 'plastic and chemistry'" (I9), or "popularising more sophisticated handicraft techniques" (I3).

Besides valuing cultural heritage as a treasure, the museum is also considered *knowledgeable* (and that is not only because of the knowledge the museum preserves). The interviewees acknowledge the knowledge and skills of museum professionals, while quite often their understanding of the knowledge needed by museum professionals to perform remains blurry. This perception is, therefore, compensated for by the hobbyists taking the expertise for granted, referring to museum work "as it usually is in museums" (I5). An honest "confession of a layman", talking about the roles of the museum, is also relevant:

> [I]n the case of textiles preservation, there can be huge differences, since, when you touch a bowl made of clay, with white gloves, nothing happens. But for this fragile textile, this is so museum-specific, I don't know what conditions it requires for preservation (I4).
>
> However, when considering an exhibition or some other communicative activity, knowledge is needed to distinguish good ideas from less good ones; as one of the interviewees states: "not all ideas are worthy of being developed" (I6).

The issue of evaluating, distinguishing or choosing may lead to *conflicts* that only the museum is capable of managing or resolving. In Table 1, several conflict situations are introduced, but probably the most topical issues for handicraft hobbyists are linked to access to collections. They are generally aware of the dilemma that exists between preserving and exhibiting fragile objects, and actively propose solutions to solve it, suggesting "making copies of objects, showing these and letting people touch them, but preserving authentic objects properly" (I8), or "digitising objects so that it wouldn't be necessary to bring things out from the repositories all the time" (I5). Yet digitising means more problems, as the lack of resources (required to deal with the vast collections) means it is necessary to prioritise, and choose between quantity (many objects digitised) and quality (lots of information attached to fewer digitised items). There is also an issue that is particularly topical for handicraft hobbyists: as they are interested in discovering new techniques, they also value information about reverse sides of pieces of furniture, garments, etc.:

> [W]hat I am missing are the wrong sides. By default, the books or photos as presented in the information system do not display wrong sides in close view. But if you want to learn some kind of technique, then the wrong side is very informative... You may want to turn a chair upside down or open the doors of a closet and have a look at what is inside (I3).

5. Estonian hobbyist crafters in relation to the Estonian National Museum

These four characteristics (the large scale, cultural values, proper knowledge and managing conflicts) are forming the identity of the ENM as an expert in the eyes of the handicraft hobbyists. In return, these traits also help to identify hobbyist crafters as museum-goers. Those traits can be first seen as in opposition to the perceived identity of the ENM.

First, compared to the vast national museum, and its collections and knowledge, hobbyist crafters perceive themselves as being rather *small and temporary*. This has an impact on their values, and, as a result, it is possible to see that handicraft hobbyists position themselves as help-seekers or users (in relation to the Estonian National Museum). Second, they express their concerns about the need to *value and popularise cultural heritage* even more, yet they feel that their own concern is not sufficient. "Proper" knowledge is the third aspect that distinguishes an individual hobbyist crafter from the museum: given their relative *lack of this knowledge,* they sometimes excuse themselves for not being au fait with matters of museology. The lack or absence of knowledge is probably one of these factors that relate the role of the ENM with the interest to actively participate in museum activities, as besides referring to little knowledge about museum work it also hints to the lack of perception of how a handicraft hobbyist could contribute to the museum. This is very vividly expressed by one of the interviewees: "I don't know how it works, therefore I cannot demand or want it... or I cannot see that it would be a problem" (I4). The end of this quote also shows that the ENM is trusted to notice and solve possible conflicts (in case there are any), since, because of their lack of knowledge, individual museum-goers (including handicraft hobbyists) *tend to distance themselves from these conflicts.*

However, there are also shared characteristics which help to contribute to commonalities and possible forms of collaboration. First, the vast collections of the museum are at least to some extent *accessed* by all participants of this intervention project (at least because the *My Favourite* contest required them to do so). In some cases they also mention visiting the collections either alone or with a group with whom they have shared interests (institutions where they work, NGOs where they are members). The interest in their native culture is shared with the ENM, as is their *interest in the wellbeing of objects* relating to their hobby or handicraft. So, despite the large scale of the museum, handicraft hobbyists also have their own "spot" related to at least a small part of collections. Being handicraft hobbyists, the interviewees *value Estonian handicraft*, and presumably their hobby is one of the main factors that helps them articulate cultural heritage as a cultural treasure.

Even though their knowledge of museum work is limited, their dedication to their hobby has in some cases formed in childhood, with the benefit of useful hints and tips from parents and grandparents. A handicraft hobbyist is, therefore, a *potentially knowledgeable person,* at least in her/his area of interest. This means that there is some acknowledgment and encouragement needed to support the specific group in their valuable interactions with the ENM. Eventually, although the general museological issues are supposed to be addressed by the museum, hobbyist crafters, as users of museum collections, publications, databases, exhibitions etc., have several ideas about how to resolve some conflicts (as was also introduced above) or about *finding new ways to collaborate.* The final part of this paper is dedicated to their suggestions about collaboration with the museum.

6. Hobbyist crafters as cultural participants in an ethnographic museum

Hobbyist crafters have proposed several ways to improve the collections, in some cases involving collaboration both on acquisition and preservation. As practical people, crafters have sometimes looked upon collections by considering both the tangible heritage preserved in repositories, and the electronic information about the collections preserved in databases, as an integral entity. Although they cherish the authenticity of objects, they also value the informative aspect of objects preserved in collections (a defect in a piece of furniture or a garment, for example, or its reverse side). Therefore, given the conflict between the need to preserve or exhibit and use fragile items, recommendations suggesting ways to add new and useful information are quite common.

They see an opportunity to contribute to the *information provision* together with the digital cataloguing of objects, "paying attention to defects, and adding instructions" (I5) on how to create a similar object. In this collaboration, handicraft hobbyists see potential for themselves in compiling the instructions (containing notes on "measurements, materials used, details and views" (I5)), with the museum professional reviewing and confirming, so that "the museum worker shouldn't have so much of a workload" (I5) (when helping crafters in research rooms).

Other ways to collaborate, through *working with the tangible heritage,* would include "restoring museum objects or crafting copies when something is very broken" (I9), or making copies to assist museum researchers (when they want to publish a textbook on some handicraft technique). "Selling handicraft to the museum" (I1) has also been considered.

A distinctive way to contribute to a museum is to send in one's *stories or interpretations*: like "the story of making my national costume" (I3), or by adding some thoughts or suggestions about collections when competing in another contest the museum might organise.

A rather specific way to collaborate, proposed by one of the interviewees, is to involve handicraft hobbyists with the required skills and knowledge in *conducting research* in fields that have been explored less thoroughly: "the work that I could definitely do would be studying items made of bones, bladders or horns. These seem to be rather unexplored" (I7).

Another area in museum work, triggering lots of ideas about collaboration between handicraft hobbyists and the museum, was related to *communication*. According to the interviewees, this area can be divided into four discrete domains: informing communities, organising exhibitions, providing courses and publishing.

7. Conclusions

While the museum, with its accumulated expertise, can be perceived as awe-inspiring and has a clear view of its relationship with the particular group of handicraft hobbyists, this scale, expertise and knowledge can in many cases also be seen as a self-construction tool. In addition to the identity-building that takes place through the relationship, these people see their role, through their self-acquired expertise as having the potential to support the museum in its endeavours.

As explored in this paper, handicraft hobbyists have proposed a rather diverse range of ways of collaborating with an ethnographic museum. Depending on the particular context, some of these suggestions may find a positive reception among museum professionals, yet some might need more time to be reconsidered or developed further. Still, it should be emphasised that the nature of these recommendations, linked to the current knowledge and museum-related identity of these handicraft hobbyists, is rather cautious, adhering very closely to previous experience. In this way, by confirming the ENM's expertise, they also re-affirm their own relationship and knowledge base through this expertise.

Acknowledgements

The authors of this paper owe special thanks to Marke Teppor, who was the main organiser of the competition "My Favourite in the Estonian National Museum's Collections", and conducted the interviews with handicraft hobbyists as part of her Master's thesis *About the Opportunities for Cultural Participation Based on the Estonian National Museum and the Craft Practitioners* in 2011. The publishing of this article was supported by the Estonian Science Foundation grant no 8006.

References

Carpentier, Nico (2007). Theoretical Frameworks for Participatory Media. In Nico Carpentier et al. (eds.). *Media Technologies and Democracy in an Enlarged Europe: The Intellectual Work of the 2007 European Media and Communication Doctoral Summer School*. Tartu: Tartu University Press, pp. 105–122.

Charmaz, Kathy (2006). *Constructing Grounded Theory: A Practical Guide Through Qualitative Analysis*. Los Angeles; London; New Delhi; Singapore: SAGE.

Desvallées, André; François Mairesse (eds.) (2010). *Key Concepts of Museology*. Armand Colin, ICOM. http://scholar.google.com/scholar?hl=en&btnG=Search&q=intitle:Key+Concepts+of+Museology#5 (accessed September 30, 2012).

Gauntlett, David (2011). *Making is Connecting: The Social Meaning of Creativity, from DIY and Knitting to YouTube and Web 2.0*. Cambridge: Polity.

Giddens, Anthony (1979). *Central Problems in Social Theory: Action, Structure and Contradiction in Social Analysis*. Berkeley and Los Angeles: University of California Press.

Giddens, Anthony (1991). *Modernity and Self-Identity: Self and Society in the Late Modern Age*. Cambridge: Polity.

Giddens, Anthony (1994). *Living in a Post-Traditional Society*. In Ulrich Beck, Anthony Giddens, Scott Lash. *Reflexive Modernization: Politics, Tradition and Aesthetics in the Modern Social Order*. Cambridge: Polity Press, pp. 56–109.

Goodnow, Katherine (2010). Introduction: Expanding the Concept of Participation. In Hanne-Lovise Skartveit, Katherine Goodnow (eds.). *Changes in Museum Practice: New Media, Refugees and Participation*. Museum of London; Berghahn Books, pp. XXV–XXXVIII.

ICOM (2006). *ICOM Code of Ethics for Museums*. Paris: International Council of Museums. Available at: http://icom.museum/fileadmin/user_upload/pdf/Codes/code2006_eng.pdf (accessed September 30, 2012).

Lasswell, Harold D. (1948). The Structure and Function of Communication in Society. In Lyman Bryson (ed.). *The Communication of Ideas*. New York: Harper, pp. 56–109.

Morrone, Adolfo (2006). *Guidelines for Measuring Cultural Participation*. Availabe at: http://www.uis.unesco.org/template/pdf/cscl/framework/CUL_particip.pdf (accessed September 30, 2012).

McQuail, Denis; Sven Windahl (1993). *Communication Models*. London: Longman.

Pruulmann-Vengerfeldt, Pille; Pille Runnel (2011). When the Museum Becomes the Message for Participating Audiences. *Communication Management Quarterly* 6 (21): 159–180.

Simon, Nina (2010). *The Participatory Museum*. Santa Cruz: Museum 2.0.

Teppor, Marke (2011). Kultuurilise osaluse võimalustest ERMi ja käsitööharrastajate näitel. Magistritöö. Master Thesis. Tartu: University of Tartu.

Tatsi, Taavi (2011). Identity Struggles of Museum Professionals: Autonomous Expertise and Audience Participation in Exhibition Production. *Journal of Ethnology and Folkloristics* 5 (2): 65–80.

Analysis of Participatory Activities in the Museums of Latvia

Linda Lotina

1. Introduction

Today's museums are competing for the attention of the visitor with other sources of entertainment, learning and dialogue. The development of alternatives increases the need to pay attention to the role of visitors and their needs in these authoritative cultural institutions. Time has changed the relationship between museums and audience from a situation in which the public had little say in museum affairs to one in which the sense of the public is an overriding factor (Combe McLean, 1994, quoted in Bandelli et al., 2009). The transformation of visitors' roles increases the importance of their opinion and decreases the gap between the authoritative museum and its visitors. This shift opens a debate about visitors as participants in cultural institutions. The concept of participation initially emerged from the political sphere but the transformation of society has raised the need to include other fields, for example culture, in the participation debate.

This paper explores the promoting and constraining factors that relate to museums in Latvia for engagement in diverse forms of participation. This study includes empirical research to record a screenshot of the practical applications of the concept of participation in the museum environment in Latvia. While being aware of a large number of possible constraints, the focus of this paper is on the post-Soviet context of Latvia and the question of collective or professional expertise, the first being more relevant as a contextual factor and the second being the key emerging theme from the empirical material.

A small Baltic state, the Republic of Latvia regained its independence in 1991 when it seceded from the Soviet Union. The re-establishment of democratic institutions included the restoration of public institutions, including museums. The aim of this study is to link the participatory activities in Latvian museums to the broader political, economic and cultural context of this post-Soviet country. The focus on the Soviet past stems from the assumption that societies facing transition processes exhibit special features that distinguish them from older democracies (Uhlin, 2009) and that context has important implications on cultural participation. The post-Soviet context means that there is an overall weakness and ineffectiveness of civil society in Eastern and Central Europe (Pettai et al., 2011) and that the experience of the formal and mandatory nature of participation under the communist regime has possibly altered societal comprehension of participation at both the institutional and individual levels.

When looking at the transformation processes in Latvia the majority of studies focus on broad economic, political, cultural and civil society issues (Dreifelds, 1996; Lejinš, Ozolina, 1997; Titma et al., 1999; Ankrava, 2003; Commission of Strategic Analysis and Advanced Social and Political Research Institute, 2005; Linden, 2008; Uhlin, 2009; Mierina, 2011). This paper focuses first on the position museums and participation in the post-Soviet context as there is a clear lack of resources with which to study museums from both the professionals' and researchers' perspectives.

This research explores how museums in Latvia put the concept of participation into practice. The analysis focuses on the main forms of participatory activities and looks at the influencing factors and attitude of museum professionals towards participation. As this article is part of a larger research project, the focus of this paper is on on-site participatory activities. The discussion about online participation is minor and is extended elsewhere (Lotina, forthcoming).

The article is structured as follows: theoretical review of the concept of participation; a description of the method; a results section providing an overview of the participatory activities in Latvian museums; a section of conclusive discussion attempting to frame the findings in the post-Soviet context.

2. The concept of participation in museums

Participation is considered a cornerstone of democratic society – the public is, through the voting process, expected to take part in the governing process. However, in times of change, the meaning of participation has been opened through a variety of different fields and hence while participation as a concept was initially political, a shift from the political to civic participation occurred (Turnšek, 2007). Carpentier writes: "The concept of participation features in surprising variety of frameworks which have been transformed through an almost infinite number of materializations" (2011: 15). He identifies the common aspects of numerous aspects of participation studies: focus on the distribution of power within society; a balance between people's inclusion in the decision-making process; differentiation of participation levels (Carpentier, 2011). Democratic theories (for example Arnstein, 1968; Pateman, 1970; Cammaerts, Carpentier 2006; Carpentier, 2011) search for balance between representation and participation and debate terms such as 'power', 'control' 'decision-making'. The focus of debate differs in the museum field. Here, museologists tend to study the relationships between interaction and participation (for example Witcomb, 2006; Heath, Lehn, 2009; Simon, 2010b) or look at the role of digital technologies in the museum's work with its audiences (for example Henning, 2006; Simon 2010a). A good example of this could be Simon (2010a) in which the terms 'power', 'decision-making' and 'control' are only mentioned superficially. For her, participation is a reinforcement of opportunities for visitors to share their own content in meaningful and appealing ways, to trust in visitors' abilities as creators, remixers, and redistributors of content and institutional acceptance of the possibility that a project can grow and change post-launch beyond the institution's original intent (Simon, 2010a). While Simon's analysis by and large ignores the power struggle so prominent in the political studies of participation, she provides a valuable analytical framework that can be adapted to organise the different participatory activities found in Latvian museums.

Simon identifies four types of participation: (1) Contributory participation – a museum is committed to help visitors and members feel like participants in the institution; (2) Collaborative participation – a museum is committed to deep partnerships with some target groups; (3) Co-creative participation – a museum is committed to support the needs of target communities whose goals align with the institutional mission; (4) Hosted participation – a museum is committed to inviting community members to feel comfortable. Each type of participation is defined by several aspects: the amount of control the institution maintains over participants; the institution's relationship with participants; selection of participants and commitment sought from participants (Simon 2010a). While the criteria differentiates the level of control and expectations of the audiences, Simon does not order these concepts as better or worse, rather, she sees them as repertoires of possibilities. At the same time, Carpentier, discussing participation of audiences in media content, identifies three basic criteria for participatory practice: (1) Decentralised power structure; (2) Trained and supported participants; (3) Participants are enabled to exercise control *and* the important role of trust and positive attitude towards participants is emphasised (Carpentier, 2003) and in this sense not all of Simon's previously mentioned four types can be seen as participatory. From that list, co-creative and hosted are modes of participation that involve the de-centralisation of power relationships and give at least some control to participants.

Simon recognises the heterogeneity of audience and distinguishes between the *audience* and *participants*. The latter are the most active members of the audience, implying that participation is not necessarily for everybody (Simon, 2010a). Individual barriers like a lack of free time, limited social capital, different lifestyles, the shortage of a good place for community, psychological barriers or lack of education (Tonn et al., 2001: 202), or lack of information literacy (Lepik, 2013) can be seen as crucial issues for any participatory project to address. Jenkins et al. mention the problem of transparency when discussing participatory culture (Jenkins et al., 2006) and linking this argument to museums, the lack of transparency in museum organisation can also discourage participation. Within museums, there is a list of less visible outputs, such as research, conservation and collection developing; therefore understanding of the mission and functions and contribution of museums is not complete.

For museums, the term *interactivity* is more inclusive and is in this context understood as a way to provide more personalised museum experience for visitors (Lewis, 2000). In the museum field participation and interactivity are blurred, especially when trying to keep in line with the progressive concept of interactivity, which, as Witcomb (2006) believes, can democratise the museum. Macdonald (2006) calls interactivity a fashionable term associated with a progressive approach.

In order to overcome the blurry boundaries of interactivity and participation I will complement Simons structuring framework with Nico Carpentier's identification of three levels when describing the relationship between people and organisations in the AIP model: (1) A for access or presence, (2) I for interaction, for example socio-communicative relationships, and also technical interaction, (3) P for a level of participation that involves co-decisions (Carpentier, 2011). AIP dimensions help in understanding the articulation of the audience role and when demonstrating how significantly participation in society can influence micro level participation in organisations. This helps to understand the importance of the post-socialist transition context.

In Carpentier's work, participation in society is fostered and supported by the participation of the media organisation or community. Here, I have adapted the model for the museum (Figure 1) in order to demonstrate how the AIP dimensions of audience activity can be linked to overall participation in society (Carpentier, 2011). Carpentier argues that while access is a first step, then participatory experience in a community, media organisation or also for instance in the museum, is an important step in preceding participation – it does not necessarily lead to active engagement in society at large. When we look at this relationship in post-Soviet countries, we can see that what Marc Morjé Howard (2003) describes as the underdevelopment of post-communist civil society can help us to analyse the role and potential of museum participation in social development. Howard (2003: 10) sees that (1) most post-communist citizens still strongly mistrust and avoid organisations, even now when participation is voluntary; (2) many of the private and informal networks that developed under communism – because of the polarisation and state control of the public sphere, as well as the shortage economy – persist today in altogether new institutional environments and serve as a disincentive for many people to join formal organisations; (3) many post-communist citizens are extremely dissatisfied with the new political and economic system, which has

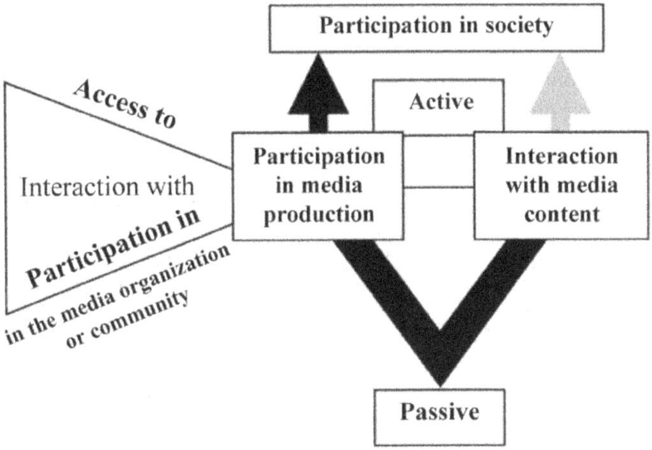

Figure 1: The micro/macro dimension of the articulation of the audience (Carpentier, 2011: 70)

not lived up to their hopes and ideas, and this disappointment has caused them to withdraw even further from public activities. As this kind of participatory environment has governed the actions and practices of museums over the long term, the question arises as to what kind of participatory activities can be sustained in the museum. Ankrava (2003) describes post-communist communication style in Latvia and among other factors they mention secrecy in decision-making, lost capability and the wish to perform teamwork. This can also become a constraint for participation, especially if power-sharing and giving up control are considered aspects of participation.

Within museums, the founding principle of a modern museum of educating the unruly masses can be difficult to harmonise with the ethos of participation in which the opinion of the masses can become a valuable input for the said museum. The values of authenticity and accuracy within the museum's internal organisation can conflict with participatory content production and recognition of the collective expertise that could be considered a necessity in valuing the participatory input of the public. Simon (2010b) illustrates this dilemma by giving the example of history museums: "Despite their support for multiple perspectives, history museums feel strongly about accuracy and authenticity. They also want to avoid stories or perspectives that reflect hateful or offensive views toward other people." Following Detel's (2005) definition of experts as individuals acquired background knowledge that gives them power to know what and why, the interpretation of labelling the museum's audience as experts or non-experts can be very much a contested divide. Pruulmann-Vengerfeldt and Runnel (2011) promote the idea that the 'average man' is an expert in his own everyday life and thus an important contributor to, for instance, an ethnographic museum. This kind of expert and non-expert division assumes a homogeneous group of museum visitors, which is by no means a valid assumption. This discussion of sharing of expertise has been taken forward especially in relation to web 2.0 technologies (for example Surowiecki, 2004; Jenkins et al., 2006; Lopez et al., 2010) with the overall conclusion that participation is a good way to involve non-expert groups whose judgment can be valid and accurate and meet the standards of museum professionals.

3. The methods

This article relies on semi-structured qualitative interviews as the main research method to reflect the plurality of museum professional opinions. 16 specialists were interviewed in the museums of Latvia during March and April 2011. Table 1 contains a description of the museums and the informants.

Table 1: Description of museums[1] and informants

Museum status	Interviewees	Notes
Local museum subjected to local municipality	Collection curator, manager of exhibition and event department, exhibition organiser	The museum displays permanent exhibitions about local history. ~ 70 000 visitors attended the museum in 2012
Regional museum. Status – municipality agency where the museum is a structure of the agency	Centre communication specialist, museum chief specialist in scientific research	The museum is established in the medieval castle and is one of the oldest museums in Latvia, situated in a popular tourism destination. ~ 96 000 visitors attended the museum in 2012
National Museum. Status – state agency	Manager of the museum education and exhibition department	The museum has the largest depository in Latvia. The museum has a disadvantageous location; ~ 45 000 visitors (including affiliates) attended the museum in 2012
Affiliate of National Museum	Department manager	Open-air museum (12-hectare-territory) displays the dwelling site from the 9th and 10th centuries. ~ 45 000 visitors attended the museum (affiliates and main organisation) in 2012

1 The description of the museums contains information from the museums' web sites and annual reports.

Public museum. Status – state agency	Chief specialist of communication department, manager of communication department, manager of scientific and education department	The museum is one of the oldest museums in Europe. It is located in Riga Old Town and is well attended both by tourists and local visitors. ~ 153 000 (including affiliates) visitors attended the museum in 2012
Affiliate of public museum	Museum specialist	Affiliate is a relatively new museum (established in 1993) and is the only museum in Riga devoted to photography. ~ 8200 visitors attended the museum in 2012
Affiliate of public museum	Director, chief specialist	The museum is located in Riga Old Town in a historical building and attracts local visitors and tourists. The Museum has managed to attract an annually increasing number of visitors. 16 000 visitors attended the museum in 2010
Affiliate of National Museum	Curator of communication, curator of communication and education	The museum is established to present Latvian professional decorative art. ~ 81 000 visitors attended the museum in 2012
Private museum	Director (also works as tour guide)	Small private museum has advantageous location close to Via Baltica highway. It is based on commercial principles. What the museum has to offer is grounded on a legendary personality, while outside the museum other leisure attractions are being developed to invite visitors to spend a longer time in the leisure complex. Visitor estimate not available

Several factors played a part in the selection of museums for qualitative interviews: covering national, regional and local museums; private and public museums; affiliates and main organisations; different thematic museums including open-air museum. As the literature review demonstrated, the notion of participation is strongly connected with the concept social media technologies, or Web 2.0. This also became a mandatory criteria for selection in which the museum had to have an internet site and an active account on at least one of the following social networking sites (SNS) – Facebook.com, Draugiem.lv[2] or Twitter.com.[3] These

2 Draugiem.lv is a Latvian social networking site established in 2004 and is the most popular social networking site in Latvia. This SNS is rather national SNS as users tend to be Latvians.

3 The same interview data were used to study online participatory practices in Latvian museums and therefore having an internet site and a social networking profile were defined as

are the most popular SNSs in Latvia: the local Latvian network Draugiem.lv engages 52% of all Latvian internet users or 32% of inhabitants (January, 2013); Facebook.com engages 33% of all Latvian internet users or 20% of inhabitants (April, 2013); Twitter.com accounts for 10% of all Latvian internet users or 6% of inhabitants (January, 2013) (Latvian Internet Association 2013). An overall review (August, 2012) of museum profiles in the most popular SNSs was carried out to identify online activity before the selection of museums. The online database *Kultūras karte* (Culture Map) provided a list of 199 accredited Latvian museums (Ministry of Culture of the Republic of Latvia; State agency "Culture Information Systems" 2006–2010),[4] of them 106 museums had no profile in any of the above mentioned SNSs, 29 museums had a registered profile that had never been active and 12 museums' profiles display only few older posts on one SNS. The rest of the museums formed the basis of a sample for interviews. From them the willingness to communicate, in this instance consent to an interview, became the deciding factor.

In most of the cases several interviews were carried out in each museum. The only exception was the private museum, as the number of staff there was so small. Questions were structured in several sections from personal to institution related ones and concerned the interviewees' professional experience, and on-site and online participatory activities in museums.

I used the museums' expected effects of the activities in order to define categories and applied content analysis to differentiate between categories. Museums expected the following groups of outcomes when organising activities: (1) Information and collective expertise collection; (2) Resources (human, material, financial resources) raising; (3) Building loyalty; (4) Attracting and educating school audience; (5) Attracting and educating general audience; (6) Engagement of stakeholders providing a range of benefits.

Balancing the analytical position between marketing and participation paradigms the data analysis is based on the understanding that everyone has the possibility to take the position of consumer and participant in various situations. The interview data was analysed with a generally inductive approach that allows research findings to emerge from the frequent, dominant or significant themes inherent in the raw data, without the restraints imposed by structured methodologies (Thomas, 2003: 2).

---

mandatory criteria by which to include a museum in the research.

4 *Kultūras karte* contains the official register of public museums and acredited private museums and by agreement non-acredited private museums.

4. Participation activities in Latvian museums

I used Simon's model to map the participatory activities in Latvian museums (Table 2). Simon describes differences among the models, however this author emphasises that none of them is more relevant or important as it "correlated with the amount of ownership, control of process, and creative output given to institutional staff members and visitors" (Simon, 2010a). Regarding participation constraints produced by the broad political and economic contexts in Latvia, the inclusive nature of a model such as Simon`s supports the selection of this theoretical framework for this study.

Table 2: Adaption of Simon's (2010a) participation models to Latvian museums

Blocks of participatory activities in museums	**Contributory participation**	**Collaborative participation**	**Co-creative participation**	**Hosted participation**	**Notes**
Information and collective expertise collection	Visitors' comments as part of exhibition content; questionnaires for general audience	Engagement of professional experts and specific groups	Engagement of professional experts and specific groups	Engagement of professional experts and specific groups	Some information collection activities are strictly marketing activities and stay outside the model. Usage of collective expertise varies from marketing purpose to high-level participation depending on situation. Online activities are used for marketing, education and information collection
Raising resources (financial, material and human)	Projects in which volunteers provide technical assistance when needed	Long-term collaborative participation with sponsors in which they keep the right to steer the direction of money usage	Projects in which the museum designs personalised work content in conformation with volunteer needs	Action called, Have Your Own Museum Item for Every Inhabitant of Cesis	Resource raising activity can stay outside the model if donor has no other impact on a process apart from donating

Blocks of participatory activities in museums	**Contributory participation**	**Collaborative participation**	**Co-creative participation**	**Hosted participation**	**Notes**
Building loyalty among participants		Special events to reward participants	Considerate volunteer programmes for retired people	Allocation of museum premises for specific target groups	Different level of participatory activities characterises loyalty building
To attract and educate school audience	Development of museum pedagogical programmes in light of pupils opinions	Collaboration with teachers preparing museum products			Even deeper partnership between these groups; museum can develop in rural territories and can reach higher level of participation
Attraction and education of general audience	Museum invites participants to create content; museum keeps decision-making rights about exhibition	Trained and consulted participants create content; museum keeps the decision-making rights to exhibit item			Participants develop content for short-term exhibitions and activity meets marketing goals as well.
Engagement of stakeholders providing range of benefits for an institution			Professionally skilled participants make decisions about the museum exhibition plan	Events organised and created by local active groups where museum acts as coordinator	Groups and individuals with professional expertise display high level participation

5. Detailed description of activities in Latvian museums

Collection of information and collective expertise. A significant amount of information collecting activities are implemented using social media. All museums widely use surveys to explore audience opinions and the needs of consumers. Museums also often read comments in SNSs and visitor books and use other available channels to access collective opinion. Museums use surveys as marketing instruments to investigate visitor opinions about souvenirs or to explore the reception of their exhibitions. Museums' interest in visitor opinions about content production is rather limited. Museums also actively use social media to educate and inform users. There are different opinions about audience activities in SNSs. Several interviewees stressed that users are passive in expressing their opinions and commenting in SNSs while only one interviewee observed that users are rather active.

The research results also show some unusual usage practices of collective expertise: the Museum of Photography used visitors' comments on photos as a part of the exhibition itself. The comments had been made many years previously when the photos were exhibited for the first time.

All the interviewees but one expressed a positive attitude towards the collective expertise of visitors, although the shortage of resources was mentioned as a hindering factor for implementation. The domination of marketing surveys as a tool to access visitor's thoughts and opinions leads to the idea that the general audience is seen as a significant provider of marketing related information, while experts are welcomed to cover a wide range of functions. The relationship with experts is built on the personal level and is rather well established.

Raising resources (financial, material and human). Latvian cultural institutions operate under conditions of limited budgets and thus raising resources is crucial for museums. Interviewees most often mentioned item donations for museum collections and recruiting volunteers. Projects aimed at acquiring object donations for museums are not always designed as participatory, although some museum professionals still see them as linking the museum and its audiences, for example donated and exhibited item can give people the feeling of participating in the development of an exhibition. In terms of participation, if and how museums develop relationships with donors is very important.

The Cesis History and Art Museum project Have Your Own Museum Item for Every Inhabitant of Cesis contains elements of donation, conservation, education and marketing. During the action, the museum invited people to select a museum item which they would take care of (which also might include taking care of restoration of the object if needed). The relationship with participants was personalised by exploring their interests, while society's knowledge of the museum's objects was increased, the museum was promoted and local inhabitants were involved in museum work. The project was designed to provide multi-faceted benefits for both sides and allocate power, control and decision-making rights to participants.

Museums actively involve volunteers to different extents, starting from exclusively inviting volunteers to assist on special occasions, strategically tying down volunteers for regular work and tailoring individual tasks, or simply by not rejecting volunteers if they come to the museum on their own initiative. Differences are seen in tasks delegated to volunteers – from highly skilled tasks to unskilled technical assistance. The content of the tasks is indicative of to what extent museums rely on participants and whether or not they are ready to share power and trust volunteers with decision-making rights.

Volunteers are usually invited and selected from specific target groups. The engagement of the groups defines their interest in museum work. Having free time and the wish to socialise (retired people); study or work in a museum related field (students, tour guides); or easy and fast contact (personal friends and relatives).

Building loyalty among participants. Some interviewees emphasise the importance of motivation and sustainable relationships with volunteers, donors, other participants. Museums organise events to honour participants or demonstrate their gratitude in other ways to make sure that participants feel appreciated and part of the museum.

Attracting and educating the school audience. Museum pedagogical programs are usually the first activity mentioned by interviewees. These programs are an effective way to attract and educate schoolchildren. Surveys of pupils' needs are used to improve the quality of the programmes.

Informative seminars for teachers and museum instigated discussions with school teachers about school children's visits to museums are other ways in which museums explore teachers' opinions, despite the fact that teachers as a target group have no direct control over the implementation of their suggestions. On the other hand, there is limited number of active schoolteachers who guide their pupils to museums and therefore museums do pay attention to these teachers' opinions not only to improve museum products for pupils but also to increase the number of visitors.

Attracting and educating the general audience. Interviewees often mentioned large-scale events as examples of participatory projects. Festivals, museum nights, traditional fairs are typical examples in this category. Two types of audience are involved general audiences and specific stakeholder groups. Events for general audiences are primarily organised to attract visitors by provide entertainment and education.

An illustrative example could be an event at the Valmiera museum, at which participants created content for the Museum Nights exhibition. Participants could receive consultation on how to develop artworks and based on this created their own art objects. The museum made a final decision on items to be exhibited for a short-term exhibition. This example of collaborative participatory activity was short-term and had no crucial impact on the museum's work, however museums in general hope this kind of event makes visitors feel closer to the museum.

Engagement of stakeholders. Projects involving stakeholders cover a wide range of groups and consequently variety of functions; museums do not hesitate to share power with these groups. The type of museum defines the stakeholders to cooperate with – for example, the National History Museum of Latvia links groups with professional expertise, such as the Society of Archaeologists.

Experts can become part of the institution and have so significant a role that the border between participants and museum workers blurs – some museum employees are members of a professional association at the same time. Their professional expertise and established relationship allow them to exert a high level of control over projects that are meaningful for both sides. Participation often occurs on occasions when external expertise form artists, scientists, etc., is needed and there is no existing cooperation between parties. However, museums display full trust in participants – for example, allowing artists to organise the opening ceremony for an exhibition.

Museums actively cooperate with youth organisations, school children and other participants from established groups like art schools. Regarding local and regional museums, active NGOs are significant partners in museum work and there are projects in which museums fully trust content creation to them. For example, when people, repressed by Soviet regime organise events to commemorate tragic events of the past in cooperation with museums, these events can be organised in cooperation with local schools or culture centres because there is not necessarily a need for specific museum expertise or usage of museum collections.

The Museum of Photography organises an annual competition for professional photographs and invites external experts to evaluate applications and to select exhibitions for the next year. The jury consists of museum specialists and art professionals from outside the museum, thus the museum gives away control of the creation of its annual exhibition plan.

During interviews, a list of other participants also emerged, such as other museums, tourism and destination marketing organisations, commercial organisations, even churches. In participatory activities that involved those groups and institutions, co-creative and hosted participation dominate. It is apparent that Latvian museums have a well-established tradition to work with such formal or informal groups and individuals.

The research results do not explicitly reflect the context of the post-socialist transition process, however during the interviews the informants' perspectives on key words such as 'trust', 'passivity', 'activity' emerge, which can be interpreted in the context of the Soviet past.

6. Conclusion: Participation in Latvian museums

This study focused upon the main types of participatory activities in Latvian museums, the attitude of museum professionals towards participants and the value of collective expertise. The interviews explicitly explored museum professionals' attitudes towards participants, however through this perspective their attitude towards the participatory approach in general become visible. The findings indicate that museum professionals are well acquainted with term *interactivity,* while *participation* is less familiar and less used. However, practical implications of the concept of participation are well visible in Latvian museums.

Howard (2003) states that the post-socialist transition context influences the relationship between the institution and participants because social and economic problems delay civil society activity. It would be inconsiderate to discuss the extent of post-Soviet impacts; however I would like to stress the significance of some participatory elements and their ratios in post-Soviet societies. Trust in other people is important for both democracy (Rozenvalds, 2009) and participatory projects (Carpentier, 2003), however only 24.8% of Latvians expressed trust in peers according to the studies of World Values Survey and the European Value Survey (Rozenvalds, 2009). Since 1990 there has been a decreasing level of membership in voluntary organisations and the trend remains negative in Latvia (Pettai et al., 2011) and participation in nongovernmental organisations did not surpass 5–6% of inhabitants (LU Sociālo un politisko pētījumu institūts 2009). The AIP model (Figure 1) displays the relationship between participation in society and participation in communities or media organisations. The context of post-socialist transition to some extent reduces society's pressure on institutions to be more open, to involve people and increase their voices.

A debate should be initiated about collective expertise as a key element of participatory projects. The attitude of Latvian museums towards collective expertise is not unequivocal and supports to some extent theoretical statements that collective public expertise is not estimated properly (Surowiecki, 2004; Jenkins et al., 2006; Lasker, Guidry, 2009). An organisation's capacity to change is shaped by how open and participative the internal flow of conversation is and how permeable its boundaries are to the diversity of external ideas (Peacock, 2008), meaning that alternative viewpoints are important for building collective expertise and reshaping an institution. Latvian museums display a range of participatory projects in collaboration with specific target groups and individual professionals who hold expertise the museum needs. Museums recognise their investments no matter whether they are local, regional or national. In all the studied institutions collective expertise produced by a general audience is welcomed, however, museums have not yet developed a system to use all the potential of the general audience even if they display a positive attitude towards collective expertise. The research findings admit that the role of museum specialists' personal attitudes towards participatory projects is as important as the much debated structure of the museum institution.

Simon has identified three institutional values that are needed in an institution to promote participation culture: desire for the input and involvement of outside participants, trust in participants' abilities, and responsiveness to participants' actions and contributions (Simon, 2010a). Museums need resources to meet these criteria; however Latvian museum professionals constantly refer to the shortage of human resources. It is understandable that museums are more open to engage in cooperation with established groups and/or groups with specific expertise. These participants display some skills and need less or no training and consultation, thus collaboration with them saves time in comparison to collaborating with previously undefined audiences. This is one of the possible explanations why participant training is still missing from many projects. The majority of museums exist in order to attract and serve as many visitors as possible, while having to compete with many other organisations for a piece of the public's leisure time (Falk, 2009), therefore for the Latvian museums which function under conditions of insufficient financing, projects with the potential to attract high visitor number could seem more attractive than developing time-consuming participatory projects.

The research results enable me to draw several conclusions regarding the nature of participation in organisations: participatory activities can involve different levels of audience engagement because of differences in motivation, attitude, etc.; the maintenance of ongoing relationship between participants and organization after a project sometimes can be significant to identify the participatory level of an activity; organization's participatory projects are interdependent, thus participation is a continuous process. Museum status – being a regional, local or national museum – does not have an impact on the number of participatory projects carried out in a particular museum. Some museums manage to develop intensive networks of participants. Regional and local museums, particularly, display several examples in which participants fulfil a range of different roles in diverse projects. On-site participatory activities dominate, although Web 2.0 tools provide good possibilities to encourage collective expertise from users. All the studied museums implement projects with different participation levels – contributory, collaborative, co-creative and hosted. Activities aim at information collection, the attraction of resources, the attraction and education of school audience and other specific target groups and general audiences. Interviewees most often mentioned contributory and collaborative activities in relation with engagement of the general audience. The projects engaging specific target groups demonstrate the highest level of participation and well-established traditions.

Regarding the post-socialist transition, studies are needed to develop an overview of low budget participatory techniques applicable in the Latvian context specifically for the general audience as far as participation with specific target groups is concerned. Debate to increase understanding of the value of participation in society is needed, however this is a broader societal debate to stimulate participatory culture.

Acknowledgements

The publishing of this article was supported by the Estonian Science Foundation grant no 8006 and the Estonian National Museum.

References

Ankrava, Sigma (2003). Postkoloniālisma sindroms un identitātes krīze Latvijā. [The syndrome of post-colonialism and identity crisis in Latvia.] *Karogs*, 1: 168–181.

Arnstein, Sherry R. (1969). A Ladder of Citizen Participation. Originally published Arnstein, Sherry, R. A Ladder of citizen participation. *JAIP*, 35 (4): 216–224. Available at: http://www.planning.org/pas/memo/2007/mar/pdf/JAPA35No4.pdf (accessed February 12, 2012).

Bandelli, Andrea; Elli A. Konijn, Jaap W. Willems (2009). The Need for Public Participation in the Governance of Science Centers. *Museum Management and Curatorship*, 24 (2): 89–104.

Comission of Strategic Analysis; Advanced Social and Political Research Institute (2005). *How Democratic is Latvia: Audit of Democracy*. Riga: The Academic Press of the University of Latvia.

Carpentier, Nico (2003). The BBC's Video Nation as a Participatory Media Practice: Signifying Everyday Life, Cultural Diversity and Participation in an Online Community. *International Journal of Cultural Studies*, 6 (4): 425–447.

Cammaerts, Bart; Nico Carpentier (2006). The Unbearable Lightness of Full Participation in a Global Context: WSIS and Civil Society Participation. In Jan Servaes, Nico Carpentier (eds.). *Towards a Sustainable Information Society Deconstructing WSIS 2006*. Bristol: Intellect, pp. 16–50.

Carpentier, Nico (2011). *Media and Participation. A Site of Ideological-democratic Struggle*. Bristol: Intellect.

Detel, Wolfgang (2005). *Foucault and Clasicc Antiquity: Power Ethics and Knowledge*. Cambridge: Cambridge University Press, pp. 6–57.

Dreifelds, Juris (1996). *Latvia in Transition*. Cambridge: Cambridge University Press.

Falk, John H. (2009). *Identity and the Museum Visitor Experience*. Walnut Creek: Left Coast Press.

Henning, Michelle (2006). New Media. In Sharon Macdonald (ed.). *A Companion to Museum Studies*. Blackwell Publishing, pp. 302–318.

Heath, Christian; Dirk vom Lehn (2009). Interactivity and Collaboration: New Forms of Participation in Museums, Galleries and Science Centres. In Ross Parry (ed.). *Museums in a Digital Age*. Oxon: Routledge, pp. 266–280.

Howard, Marc M. (2003). *The Weakness of Civil Society in Post-Communist Europe*. Cambridge: Cambridge University Press.

Jenkins, Henry; Katie Clinton, Ravi Purushotma, Alice J. Robison, Margaret Weigel (2006). *Confronting the Challenges of Participatory Culture: Media Education for the 21st Century*. Chicago: MacArthur Foundation. Available at: http://digitallearning.macfound.org/atf/cf/%7B7E45C7E0-A3E0-4B89-AC9C-E807E1B0AE4E%7D/JENKINS_WHITE_PAPER.PDF (accessed February 12, 2012).

Lasker, Roz Diane; John A. Guidry (2009). *Engaging the Community in Decision Making: Case Studies Tracking Participation, Voice and Influence*. Jefferson: McFarland.

Latvian Internet Association (2012). Statistics. Available at: http://www.lia.lv/statistika/ (accessed February 12, 2013).

Lejiņš, Atis; Žaneta Ozoliņa (1997). *Small States in a Turbulent Environment: The Baltic Perspective*. Riga: Latvian Institute of International Affairs. Available at: http://liia.lv/site/attachments/17/01/2012/Small_States_in.pdf (accessed February 12, 2012).

Lepik, Krista (2013). *Governmentality and Cultural Participation in Estonian Public Knowledge Institutions*. Tartu: University of Tartu Press.

Linden, Tove (2008). *Explaining Civil Society Core Activism in Post-Soviet Latvia*. Stockholm University. Available at: http://www.diva-portal.org/smash/get/diva2:198259/FULLTEXT01.pdf (accessed February 12, 2012).

Lewis, Jelena (2000). Interactive Museum. In *TechKnowLogia*, September/October: 52–53. Available at: http://www.techknowlogia.org/TKL_Articles/PDF/182.pdf (accessed February 12, 2012).

Lopez, Ximena; Ilaria Margapoti, Roberto Maragliano, Giuseppe Bov (2010). Digital Heritage. The Presence of Web 2.0 Tools on Museum Websites: a Comparative a Comparative Study between England, France, Spain, Italy, and the USA. *Museum Management and Curatorship*, 25 (2): 235–249.

Mieriņa, Inta (2011). *Political Participation and Development of Political Attitudes in Post-communist Countries*. Ph.D. Thesis, University of Latvia. Available at: http://szf.lu.lv/files/petnieciba/promocijas_darbi/Disertacija%20-%20Mierina%20-%20Development%20of%20Political%20Attitudes.pdf (accessed February 20, 2012).

Ministry of Culture of the Republic of Latvia; State agency “Culture Information Systems” (2006–2010). Kulturaskarte.lv. Museums. Available at: http://www.kulturaskarte.lv/lv (accessed February 28, 2013).

Pateman, Carole (1970). *Participation and Democratic Theory.* Cambridge: Cambridge University Press.

Peacock, Darren (2008). Making Ways for Change: Museums, Disruptive Technologies and Organisational Change. *Museum Management and Curatorship*, 23 (4): 333–351.

Pettai, Vello; Daunis Auers, Ainė Ramonaitė (2011). Political development. In Marju Lauristin (ed.). *Estonian Human Development Report 2010/2011. Baltic Way(s) of Human Development: Twenty Years On.* Tallinn: Estonian Cooperation Assembly, pp. 144–164.

Pruulmann-Vengerfeldt, Pille; Pille Runnel (2011). *Increasing the Usability of the Museum: Four Studies*. Available at: http://www.idc.ul.ie/techmuseums11/paper/paper21.pdf (accessed February 12, 2012).

Rozenvalds, Juris (2009). Latvijas demokrātija starp "eiropeizāciju" un postpadomju telpas mantojumu. [The democracy of Latvia between "europeization" and the heritage of post-soviet space.] In Valts Kalniņš (ed.). *Pārskats par korupciju un pretkorupcijas politiku Latvijā.* [Review of corruption and anticorruption policy in Latvia]. Riga: Soros Foundation – Latvia, pp. 18–30. Available at: http://politika.lv/article_files/1887/original/Korupc_09_1_bez_vaka.pusg.pdf?1339437240 (accessed February 28, 2013).

Rozenvalds, Juris (2010). Padomju mantojums un integrācijas politikas attīstība kopš neatkarības atjaunošanas [Soviet heritage and integration policy development since renewal of the independence]. In Nils Muižnieks (ed.). *Cik integrēta ir Latvijas sabiedrība? Sasniegumu, neveiksmju un izaicinājumu audits*. [How integrated is society of Latvia? Auditing achievments, failures and challenges.] Riga: Latvijas University of Latvia, pp. 33–60. Available at: http://szf.lu.lv/files/petnieciba/publikacijas/Integrac_aud_latv_val_jpg.pdf (accessed February 28, 2013).

Simon, Nina (2007). Discourse in the Blogosphere. *Museums & Social Issues*, 2 (2): 257–274. Available at: http://www.museumtwo.com/publications/MSI_257-274_simon.pdf (accessed February 12, 2012).

Simon, Nina (2010a). *The Participatory Museum*. Available at: http://www.participatorymuseum.org/read/.

Simon, Nina (2010b). *How Different Types of Museums Approach Participation* Available at: http://museumtwo.blogspot.com/2010/02/how-different-types-of-museums-approach.html (accessed February 12, 2012).

State Agency "Culture Information Systems" (2007). Central Catalogue of National Museum' Collection. Available http://www.nmkk.lv/PubliskaSakumlapa.aspx (accessed February 28, 2013).

Surowiecki, James (2004). *The Wisdom of Crowds: Why the Many are Smarter than the Few and how the Collective Wisdom Shapes Business, Economies, Societies and Nations*. New York: Doubleday.

Thomas, David R. (2006). A General Inductive Approach for Analyzing Qualitative Evaluation Data. *American Evaluation Journal*, 27 (2): 237–246.

Titma, Mikk; Nancy B. Tuma, Brian D. Silver (1999). *Winners And Losers In The Post-Communist Transition: New Evidence From Latvia*. Washington: The National Council for Eurasian and East European Research. http://www.ucis.pitt.edu/nceeer/1999-813-01-Titma.pdf (accessed February 12, 2012).

Tonn, Bruce E.; Percides Zambrano, Sheila Moore (2001). Community Networks or Networked Communities? *Social Science Computer Review*, 19: 201–212.

Turnšek, Maja (2007). New Media, Transformations of Participation, and the Problem of Publicness. In Nico Carpentier, Pille Prullman-Vengerfeldt, Kaarle Nordenstreng, Maren Hartmann, Peeter Vihalemm, Hannu Nieminen, Tobias Olsson (eds.) *Media Technologies and Democracy in Enlarged Europe: The Intellectual Work of the 2007 European Media and Communication Doctoral Summer School*. Tartu: Tartu University Press, pp. 187–200.

Uhlin, Anders (2009). Which Characteristics of Civil Society Organizations Support what Aspects of Democracy? Evidence from Post-communist Latvia. *International Political Science Review*, 30 (3): 271–295.

Witcomb, Andrea (2006). Interactivity: Thinking Beyond. In Sharon Macdonald (ed.). *A Companion to Museum Studies*. Oxford: Wiley-Blackwell, pp. 353–361.

[illegible]

[illegible] Networks [illegible]

[illegible]

STRUGGLES OF MUSEUM PROFESSIONAL

Facing the Death of the Author. Cultural Professional's Identity Work and the Fantasies of Control

Nico Carpentier

1. Introduction

Barthes' *Image Music Text* (1984) contains the seminal essay *The Death of the Author*, which pointed to the convergence between the producers and receivers of discourses at the level of interpretation. The death of the modernist Author was a metaphor, not be taken literally, implying that there was no privileged vantage point that fixed the interpretation of a text. But it also referred to structural power changes in society, where members of cultural elites could no longer claim control over their writings. 'Ordinary' readers became (seen as) more and more capable of producing their own interpretations, which might structurally diverge from the intentions of the Author. As we have more recently witnessed an increased convergence between the producers and receivers of discourses at the level of the production process, we could say the Author died a second time. The old Author is no longer solely in control of the production process, as the 'produser' (e.g. Bruns, 2007) has overcome the rigid separations between both categories. Again, this is seen as a major step towards the democratization of our cultural realms.

There are a number of problems with this type of argument. First of all, the argument tends towards an individualized interpretation of the social, which leads to a downplaying of societal structures, including the importance of organizational structures in providing cultural elites with safe havens, and the importance of discursive structures like professional identities and audience identities. Obviously, these structures are interdependent, as institutions act as discursive machineries, producing identities, and professional identities are driving forces for the functioning and legitimization of cultural institutions. These structural components make the Author more resistant than it seems. S/he has indeed found shelter in a series of organizations and institutions, protected by their professional structures and organizational cultures that provide networks of support and resources. The Author is also resistant at the cultural-discursive level, as the contemporary subject positions related to the Author (or the many cultural professionals[1]) turn out to be more rigid than expected (and sometimes desired). In other words, cultural

1 Cultural professionals are seen in this article as people that are professionally active in cultural institutions and that are involved in processes of cultural production. From this perspective, they are condensations of the Author discourse(s).

professionals' identities remain embedded within hegemonic discourses on for instance management, autonomy and expertise.

But this is not the only problem with the death-of-the-Author argument. Especially in the 1990s and 2000s we have witnessed a strong revival of this argument, connected to the changes in the communicational landscapes. The popularization of the internet, with all its potentials for interaction and participation fed into the cultural democratization argument, combined with the belief that these changes were new and driven by Information and Communication Technologies (ICTs). This claim for novelty is highly problematic, as it tends to ignore the history of cultural participation, which stretches out much further than utopian ICT theories want us to believe. Articulating ICTs as the driving force of the social is equally problematic, as societal changes are multidimensional and highly context-dependant. Moreover, ICTs are not the only sites of meaning production, as the social consists of a multiplicity of discursive machineries.

This article wants to discuss the role of Author-related discursive structures (like subject positions) in participatory processes within the cultural realm, (partially) focussing on museum studies examples, which provides unusual but very rich and relevant case study material for media scholars interested in audiences and professionals. Arguably, the cultural-democratic discourses and practices that have been circulating for a considerable amount of time have required cultural professionals to develop specific strategies – termed identity work here – to deal with these discourses. The theoretical assumption in this text is that identities (or subject positions) are not stable or homogenous, but contingent and diverse, and fed by social fantasies. This assumption (supported by culturalist identity theory and psychoanalytic theory) allows analyzing how the cultural professional has been articulated through a series of contemporary fantasies. This article will first focus on the resistant modernist fantasy of the cultural professional as Author, which sometimes takes an antagonist/oppositional position towards the audience, and then move on to its still modernist counterweight, the democratic-populist fantasy of the death of the Author. The strong disconnection of these fantasies with the present-day cultural-democratic configuration, characterized by a more post/late/liquidmodernist logic, allows for the articulation of a third fantasy, which foregrounds participation, but reframes it to increase its alignment with this post/late/liquid-modernist culture through the recognition of difference and conflict by placing it within an agonist framework.

2. Identities at work

As the notion of identity carries many different meanings, it is instrumental to start by explaining how I use the concept here. In relation to the two major theoretical strands that define identity, namely the more psychological strand (personal identity) and the more sociocultural strand (social or cultural identity), this text aligns itself with the second strand. More specifically, identity is seen as a discursive structure that provides meaning to objects, individual and collective agents. From this perspective, the social is characterized by a multitude of circulating identities, contested and contestable, that offer subjects opportunities for identification (which creates the link with the more psychological approaches) and provide them with the building blocks of their subjectivities. Support for this position can be found in Sayyid and Zac's (1998: 263) approach, when they write that identity is to be defined in two related ways. First, identity is "*t*he unity of any object or subject". This definition is in line with Fuss' (1989: ix) definition of identity as "the 'whatness' of a given entity". A second component of the definition of identity comes into play when the concept is applied to the way in which social agents are identified and/or identify themselves within a certain discourse. Examples Sayyid and Zac (1998: 263) give in this context are "workers, women, atheists, British."

Laclau and Mouffe call this last component of identity a subject position (i.e., the result of the positioning of subjects within a discursive structure), which will be used in this analysis to describe the discursive positionings of actors. An important characteristic of the subject position concept is that it emphasizes the role of discursive structures to provide people with positions within the social, but simultaneously allows space for the contingent articulation of these positionings:

> Whenever we use the category of 'subject' in this text, we will do so in the sense of 'subject positions' within a discursive structure. Subjects cannot, therefore, be the origin of social relations – not even in the limited sense of being endowed with powers that render an experience possible – as all 'experience' depends on precise discursive conditions of possibility. (Laclau, Mouffe 1985: 115)

In order words, Laclau and Mouffe's definition implies neither a structuralist nor a voluntarist position. Although they endorse Althusser's critique of the autonomous and self-transparent subject (a voluntarist position), they vehemently reject Althusser's economic determinism (a structuralist position), because in their view this aspect of Althusser's theory leads to a "new variant of essentialism" (Laclau, Mouffe 1985: 98). However, Laclau and Mouffe's rejection of this aspect of Althusser's work does not keep them from borrowing from him the originally Freudian concept of overdetermination, although not without altering its meaning:

> *S*ociety and social agents lack any essence, and their regularities merely consist of the relative and precarious forms of fixation which accompany the establishment of a certain order. This [Althusser's] analysis seemed to open up the possibility of elaborating

> a new concept of articulation, which would start from the overdetermined character of social relations. But this did not occur. (Laclau, Mouffe 1985: 98)

The notion of overdetermination is one of the strategies that Laclau and Mouffe use to emphasize the contingency of the social and of identities. This contingency can already be found at the heart of their discourse theory, namely when they are discussing the nature of discursive structures (including identities and subject positions), the importance of articulation, the floating of signifiers and the infinitude of the field of discursivity. A discourse is seen as a structured entity that articulates different elements, whose meaning is altered by the process of articulation itself. Inspired by early semiology, Laclau and Mouffe (1985: 106) claim that "all identity is relational", which implies the establishment of relationships of inclusion and exclusion, but also a process of modification. This becomes clear in their definition of articulation, which is seen as a "practice establishing a relation among elements such that their identity is modified as a result of the articulatory practice" (ibid.: 105). Contingency originates from the specificity of the articulated elements (where some elements become articulated in a discourse, and others are not – they remain available in the field of discursivity), from the process of articulation and the specificity of the combination of elements, and from the possibility of re-articulation (where new elements become articulated or old elements become dis-articulated, which affects the entire discourse).

But also in Laclau and Mouffe's political identity theory (which builds upon their discourse theory in the strict sense – see Carpentier, Spinoy, 2008) contingency features prominently, as the political is seen as a site of conflict, antagonism and struggle for hegemony (see also Mouffe (2005) for an elaborate argumentation). Although their political identity theory focuses more on the attempted stabilizations of the social through hegemonizing processes, they still base their theory on an ontology of contingency where hegemony can never be total. As Mouffe (2005: 18) writes:

> Every hegemonic order is susceptible of being challenged by counterhegemonic practices, i.e., practices which will attempt to disarticulate the existing order so as to install other forms of hegemony.

Also the actual process of establishing a hegemonic social imaginary presupposes societal contingency. This struggle for hegemony takes place in "a field criss-crossed by antagonisms" (Laclau, Mouffe, 1985: 135), where different sets of identities are aligned into a hegemonic project[2] and opposed to another negative

2 This happens through the so-called logic of equivalence, however without totally eliminating their differences: A chain of equivalence can weaken, but not domesticate differences (Laclau, 2005: 79).

identity, a constitutive outside. Through the interplay between antagonistic identities, these identities become constructed and can (in some cases) gain dominance. But Laclau and Mouffe's negative-relationalist approach to identity also allows them to show the limits of the formative capacity of antagonism (in constructing identities), as the presence of the 'other' identity remains a necessary component in the identity construction process. This means that identity can never be fully developed and foreclosed: "The presence of the Other prevents me from being totally myself" (Laclau, Mouffe, 1985: 125). Antagonistic identities try to (discursively) eliminate each other but simultaneously need each other as each other's outsides.

Despite Laclau and Mouffe's careful positioning of the subject between structuralism and voluntarism, Žižek critiqued their reduction of the subject to its subject positions. In an essay published in Laclau's *New Reflections on the Revolution of Our Time*, Žižek (1990: 250) explained this reduction as "an effect of the fact that Laclau and Mouffe had progressed too quickly" and did not manage to combine the "radical breakthrough" at the level of the concept of antagonism with an equally well elaborated theory of the subject. This criticism has led especially Laclau to acknowledge "the importance of an understanding of subjectivity in terms of the subject-as-lack" (Glynos, Stavrakakis, 2004: 202) Although in *Hegemony and Socialist Strategy* (Laclau, Mouffe, 1985) identities were already seen as a fusion of a multiplicity of identities, where the overdetermined presence of some identities in others prevents their closure, Laclau's later work more clearly distinguishes between subject and subjectivation, identity, and identification. The impossibility of the multiplicity of identities to fill the constitutive lack of the subject prevents their full and complete constitution because of the inevitable distance between the obtained identity and the subject, and because of the (always possible) subversion of that identity by other identities. In Laclau's (1990: 60) own words: "the identification never reaches the point of full identity". Or as Sayyid and Zac (1998: 263) put it: "the subject is always something more than its identity". As Torfing (1999: 150) illustrated, there are many possible points of identification:

> A student who is expelled from the university might seek to restore the full identity she never had by becoming either a militant who rebels against the 'system', the perfect mother for her two children, or an independent artist who cares nothing for formal education.

Precisely the contingency of identities and the failure to reach a fully constituted identity creates the space for subjectivity, agency, freedom, and the particularity of human behavior:

> The freedom thus won in relation to the structure is therefore a traumatic fact initially: I am condemned to be free, not because I have no structural identity as the existentialists assert, but because I have a failed structural identity. This means that the subject is partially self-determined. However, as this self determination is not the expression of what the subject already is but the result of the lack of its being instead, self-determination can only proceed though processes of identification. (Laclau, 1990: 44)

The self-determination that Laclau mentions generates space for subjects to become actively involved in the identity construction process, working with the building blocks that are available within the social, (re-)articulating and performing them, struggling against them and adopting them. Identity politics (and the politics of identity – see Hall, 1989) is for instance very much based on the political agency of those engaged in the deconstruction of dominant identities. Another concept that refers to the active role of subjects in dealing with their identities, is identity work. This concept – originally used at a more individual level (see Snow, Anderson, 1987) but later applied to collective identities and subject positions (see, for example, Reger, Myers, Einwohner, 2008) – captures the discursive efforts that people have undertaken in order to (re)construct and maintain their identities.

This self-determination is of course not unlimited. As Laclau (1990: 44) argues, "selfdetermination can only proceed though processes of identification", which generates the connection with discursive structures (or subject positions) which are outside the subject itself. At the same time, there is a strong desire for the wholeness of identities and the harmonious resolution of social antagonisms, although this wholeness and harmony is structurally lacking. If we turn to a Lacanian perspective, we can see that desire is conceptualised exactly through a relation with a lack (and not as a relation to an object). What causes the desire is exactly the lack, the incompleteness of identity, which lies at the core of all subjectivity (Lacan, 1991: 139; Kirshner, 2005: 83). Subjects crave for fully-constituted identities, but these can never be realized. The lack can never be filled; the desire can never be satisfied. Desire is the "lack of being whereby the being exists" (Lacan, 1988: 223) which turns it into an endless unconscious driving force. The mechanism that allows dealing with this structural inability and the frustration it generates, is fantasy, as fantasy provides us with hope and protection (Lacan, 1979: 41). Fantasy provides the subject with the (imaginary) frames that conceal and promise to overcome the lack (Lacan, 1994: 119–120); in this way, fantasy functions as "*t*he support that gives consistency to what we call 'reality'" (Žižek, 1995: 44). Nevertheless, this ultimate victory remains out of reach, and eventually all fantasies become again frustrated and their limits visible, showing the contingency of identity and the social.

3. The subject position of the cultural professionals

When turning to the relations between cultural professionals and audience members in cultural institutions, also their interaction is structured by their identities (or subject positions). Embedded within a societal context, these identities are not completely rigid, but can become re-articulated over time (and space). Simultaneously, radical re-significations are rare, and traces of older articulations remain present in contemporary subject positions. One example here is the postmodern itself which still contains (traces of) the modern, as Lyotard (1984: 78) puts it: "A work can become modern only if it is first postmodern. Postmodernism thus understood is not modernism at its end but in the nascent state, and this state is constant". A similar argument can be made about the modern cultural professional, who is still articulated as Author.

In the case of the modernist articulation of the cultural professional – the cultural professional as Author – the use of the word 'profession' provides us with a series of meanings that are attributed to this (articulation of this) subject position. Using McQuail's (2008: 53) helpful list of characteristics of the professional (but also work more focussed on the media professional (Carpentier, 2005; Deuze, 2005)), we can distinguish a series of signifiers that construct the subject position of the modernist cultural professional, in an oppositional or sometimes antagonistic relationship with the identity of the audience. Together they form an equivalential chain of particularities that construct the identity of the modernist cultural professional.

A first basic element is the notion of expertise, which is acquired through training and education. Expertise is based on a combination of knowledge and skills, which structures and legitimizes the decisions that allow for cultural production, but that also distinguish the cultural professional from the audience of the cultural production. In a more traditional articulation, these knowledges and skills would encompass contextual knowledge (for instance about the field, its history and its actors) and object knowledge (for instance the canonical meanings of the objects to be displayed), but also the skills to use technologies of display to translate these knowledges into spatial orderings and secondary texts. But in more market-driven environments, these knowledges and skills would be complemented and sometimes replaced by market-related knowledges (for instance about the potential visitors and target groups) and management skills. Quite often this expertise is based on what Bourdieu (2000) called legitimate knowledge (and skills), and can be seen as a way to impose a legitimate vision on the world, while other types of knowledge (and skills), like the situated knowledges circulating in communities, are facing the permanent risk of being discredited.

A second element of the professional subject position is the public service that is provided to both specific audiences and society in general. Cultural professionals are articulated as transcending self-interest and commercial and institutional interests. As care-takers of cultural heritage and enablers of cultural production and education, they act out of a vocation or calling, maintain a certain degree of detachment and impartiality, and provide a cultural experience to an audience which is attributed societal ('public') value that cannot be reduced to its exchange value. This positioning also affects the audience identity, which often finds itself articulated as passively receiving the service provided to them (although the activation of the audience could be seen as a form of public service).

A third element is linked to the concept of ethics, which is in turn connected to notions of truth, authenticity, integrity and honesty. A wide range of ethical principles can be invoked: In their discussion of media ethics, Christians, Rotzoll and Fackler (1991) for instance refer to Artistotles' golden mean, the Kantian categorical Imperative, Mill's principle of utility, Rawl's veil of ignorance, and the Judeo-Christian's persons as ends principle. Whatever framework is preferred, ethical behavior is seen as an intrinsic part of the identity of the cultural professional. This need for ethical behavior is not exclusively related to the outcomes of the cultural production process (for instance requiring truth-telling) but also impacts on the process itself (for instance requiring the proper treatment of 'outside' actors). Again, ethics generate a difference between cultural professional and audience identities, as audience members are not bound by the same ethical principles (although they are for instance bound by codes of conduct).

The public service and ethical behavior generate a logic of difference which legitimizes a certain degree of autonomy to the cultural professional, which is further strengthened by the identity's link with the epistemological framework of expertise. Expertise is still very much seen as an individualized activity, which requires the protection against 'outside' intervention to come to fruition. Cultural production has often been regarded as a freespace where the colonizing forces of the market and state would not manage to penetrate (in its entirety), which legitimized the need – some would say the myth – for autonomy as a key identificatory signifier. But also in more commodified cultural environments, autonomy remains to play an important role, as the capitalist enterprise still structurally privileges individualized expertise that is autonomously deployed. Finally, the audience takes on a specific position here, as it is seen to pose a potential threat towards the autonomy of the cultural professional, as 'unwarranted claims' from individuals, organizations, stakeholders or communities might attempt to affect the cultural professional's activities.

The fifth element is the institutional embeddedment of cultural professionals, which is often translated into a relation of employment. Cultural professionals are rarely articulated as amateurs, although they sometimes can be employed as free-lancers. However weak (and often problematic) the link of employment is in the latter case, it still exists and structures the identity of the cultural professional. Here employment creates an important indicator for this professional identity as it regulates the access to the professional system which supports this identity. Moreover, these institutional environments provide cultural professionals with support systems but also with the presence of peers, who perform and protect the professional culture. Integrated into networks of peers, cultural professionals can define themselves as members of a professional/intellectual/artistic elite, which is articulated as different from (for instance) audience members, who through this oppositional logic become positioned as 'ordinary'.

This brings us to our last element, the deployment of management and power. Cultural professionals are often placed in a hierarchically structured entity and attributed specific responsibilities for the professional production of specific cultural products. This responsibility is complemented by the notion of psychological property (Wilpert 1991). To realize the professional goals, cultural professionals can make use of the production facilities that are owned (in the strictly legal sense of the word) by the media organization. Wilpert's (1991) theory of psychological appropriation provides support for the thesis that the control over these production facilities leads to a sense of property. It is precisely this combination of responsibility, (psychological) property and authorship that supports the articulation of the cultural professional as a manager of a diversity of resources, from technology via content and objects to people. I should of course be careful not to attribute absolute power to cultural professionals (eliminating the possibility of resistance of those who are affected, but also the influence of the organizations' hierarchy), but cultural production often entails the management of audiences' bodies and the targeted exposure of audiences' minds to carefully selected meanings.

4. The opening up of the cultural realms

One can wonder whether this modernist articulation of the cultural professional still has some connection to our everyday worlds. The contemporary context of postmodernity (or late / liquid modernity) unavoidably increases the levels of hybridity and liquidity in the social configuration. Processes like ideological fragmentation and cultural amalgamation (or the end of the 'great divide' between low and high culture – see Huysen, 1986), but also of detraditionalization, individualization, globalization and commercialization (see, for example, Krotz, 2007) have had a structural impact on contemporary societies and have affected the circulating power relations, in the political realm but also in the cultural realm. The democratic revolution has not only increased popular participation in institutionalized politics (at least when looking on the long term), but opened up and decentralized different other societal fields.

One field that is often attributed a key role is the field of so-called 'new' media, although care should be taken not to fall into a communicational reductionist trap. Of course, the arrival of a new generation of media technologies did impact on democracy, participation, and the media system, and did put pressure on the Author (mainly the media professional). In the 21th century, not only ordinary users but also civil society organizations (van de Donk et al., 2004; Cammaerts, 2005) are more enabled or empowered to avoid the mediating role of the 'old' media organizations, to publish their material (almost) directly on the web, and to establish communicative networks that (often) support more decentralized models of democracy. There is a potentially beneficial increase in information, which challenges the "existing political hierarchy's monopoly on powerful communications media" (Rheingold, 1993: 14), might result in the strengthening of social capital and civil society (Friedland, 1996), and might even open-up new public spheres, or "global electronic agora[s]" (Castells, 2001: 138). But we should keep in mind that the role of ICTs to deepen the democratic process is contextdependant.

ICTs can have many different applications, and can be used in many different constellations. In other words, ICTs are not inherently democratic, although some ICTs might have characteristics that can facilitate more democratic-participatory usages. ICTs remain firmly embedded within their societal contexts, where we can see the political, social, cultural and technological interlock in a dynamical process, feeding into societal change or sedimentation, into processes of hegemonization or resistance, into historical continuities or ruptures.

The fetishization of media technologies can be avoided by looking at a series of other fields. Lyotard (1984) pointed – quite some time ago – to the changes in field of science: While science managed to mobilize a self-legitimizing metanarrative in the 18th and 19th century, the fracturing and dismantling of discourses ended science's position of taken-for-grantedness in the 20th century. These problems of self-legitimization not only affect science, but also other expert fields, as Lyotard (1984: 14) explains:

> The ruling class is and will continue to be the class of decision makers [...] [but] the old poles of attraction represented by nation-states, parties, professions, institutions, and historical traditions are losing their attraction.

Using another perspective, Beck, Giddens and Lash (1994) refer to the contemporary configuration as a reflexive modernity, based on the realization of a wide range of democratic ideals, and the shift from emancipatory and centralized politics to life politics and/or subpolitics. These life / subpolitical issues are both global and part of everyday life, and still provide expert systems ("scientists and professionals" (Lash, 1994: 198)) with significant roles. As Lash (ibid.) remarks, these expert systems are "affecting everyday life", but they are "now open to democratic debate and contestation from the lay population". Beck, Giddens and Lash (1994) locate a prime source of social change in these expert systems, as they might constitute new public spheres. At the same time, Foucault's governmentality model produces a slightly less optimistic perspective on the social, where ever more sophisticated disciplinary and post-disciplinary power plays work through individualized freedom to still generate (and legitimize) societal control.

Also the institutions of display and conservation – the museum – became implicated in the debates about participation, as a series of museum theorists started to advocate a new museology or new museum theory. One foundational text was Vergo's (1989a) anthology, appropriately entitled *The New Museology*, in which he and a number of authors advocated a reconfiguration of our ways of looking at the museum. In his introduction, Vergo (1989b: 3) refers to the dissatisfaction with the 'old' museology, which focussed too much on museum methods, and was not reflexive enough about the museum's purposes and identities. In the same introduction, Vergo also distanced himself from claiming ultimate novelty and exclusivity,[3] and mono-perspectivism. Within this diverse collection of articles, a number of authors (plead to) rethink the museum's relation to the visitor, and the power imbalances that characterise that relationship. For instance, Merriman (1989: 167–168) – drawing heavily on Bourdieu's (1984) concept of distinction – concludes that

> … the action of museums in contemporary culture is to divide society into those who have the 'competence' to perceive museum visiting as a worthwhile leisure opportunity, and those who do not.

Wright (1989: 148) takes a similar position:

> The present fiction in museums – that every visitor is equally motivated, equipped, and enabled 'to experience art directly' – should be abandoned. It is patronising, humiliating in practice, and inaccurate.

Secondly, also the political nature of the museum and its functioning as a discursive machinery is thematized. Especially Greenhalgh's (1989: 96) chapter on international exhibitions offers a strong case, where he shows how these exhibitions "recognized the socio-political climate of their time and how they responded to it".

3 See Halpin (1997) for a brief historical analysis of earlier museum (theory) reform projects.

In later publications on new museology / new museum theory, this emphasis on representation, the political and power is deepened, and combined with a more explicit agenda for social and cultural change. Critiques on the elitism, exclusionary practices and monovocality of museums (Ross, 2004) form the basis of a museum reform project that aims for "the transformation of the museum from a site of worship and awe to one of discourse and critical reflection that is committed to examining unsettling histories with sensitivity to all parties" (Marstine, 2006: 5). Secondly, also the emphasis on the inclusion of the museums' communities is continued, witness Marstine's (2006: 5) plea for a museum that "is transparent in its decision-making and willing to share power". Through this strong emphasis on inclusion and power, the notion of audience participation is brought into the debate again, for instance through the recognition that visitors and communities also have cultural expertise, as Halpin (1997: 56) writes:

> The new or critical museology about which I am speaking might be a useful museology in service to a community, instead of the state and the élite. A museology practised by named, committed and creative professionals who know that people other than themselves are also cultural experts.

Readers like *Cultural Diversity. Developing Museum Audience in Britain* (Hooper-Greenwill, 1997) and *Museums, Society, Inequality* (Sandell, 2002) focus strongly on the importance of inclusionary practices, combined with the provision of series of examples. One example is Hemming's (1997) chapter in the first reader, which has the (rather telling) title *Audience Participation: Working with Local People at the Geffrye Museum*. In this chapter, Hemming discusses the exhibition *Chinese Homes: Chinese Traditions in English Homes*, which ran for three months in the Geffrye museum in Hackney (London), in combination with the educational courses organized by the museum for different groups of people within the community. Through the collaboration with a Chinese Community Centre, members of the Chinese community were involved in the construction of the *Chinese Homes* exhibition, by combining group discussions on content (and access to preparatory meetings) with oral history approaches. In his non-celebratory process evaluation, Hemming (1997: 176) points to the problems related to language, resources and time, but also emphasizes the importance of audience participation:

> Involving the community in making decisions does take time, but also the will to make it happen. However, if the museum had tried to impose its own narrative on the exhibition without the consultation process, the results would have been disastrous. The chances are that the exhibition would have alienated the Chinese community and been a rather shallow attempt to portray their culture.

5. Cultural professionals, fantasy and their identity work

These transformations have put pressure on cultural professional subject positions, requiring them to perform additional identity work. In the case of cultural professionals this identity work implies the development of coping strategies to deal with – amongst other issues – the increase of audience, visitor, reader, and spectator power to interpret cultural products on their own terms, to demand to become involved in the machineries of cultural production and to produce cultural artefacts themselves.

The modernist articulation of the cultural professional has not fully disappeared, but it has been highly problematized in contemporary societies. Nevertheless, cultural institutions still provide shelter to this articulation, which is based on the fantasy of full control and management. In some cases this leads to nostalgia, where the complexities of fluidity and hybridity are mourned over and the return to a more straightforward past with 'clear' subject positions is desired for. In other cases antagonistic identity strategies are applied, whereas these audience members, visitors, readers, and spectators are defined as others, sometimes even 'enemies'. Through these dichotomising articulatory processes, ordinary people are constructed as a homogeneous mass, and detached from social structures (like civil society or communities). Their everyday life knowledges are discarded as irrelevant and illegitimate. They are deemed to lack any expertise, and in dire need for education. Their behaviour is considered to be uncivilized and a potential threat (for instance to the cultural objects on display), which necessitates the deployment of sophisticated management techniques. For instance Macdonald's (2002: 160) ethnography of the Science museum (in London, UK) provides some nice examples of antagonistic staff members' rhetoric on the audience:

> In everyday talk in the Museum it was fairly common for visitors to be referred to as problems, as "in the way", as disruptive and as "stupid". [...] For many curators that visitors might not understand certain Museum-imparted information was evidence of visitor ignorance. Stories would circulate about visitors who had completely misunderstood exhibits in amusing ways – perhaps trying to look into the wrong part of an interactive [display] or confusing an effect with a cause. Visitors were also sometimes depicted as deviants, especially as vandals.

In other cases, more benevolent (but not necessarily less problematic) discourses are used to construct a difference between the cultural professional and the societal groups they aim to serve. Here we can for instance mention the strategy of respectful detachment, where the otherness is acknowledged and the other is respected but no attempt for communication or interaction (let alone participation) is initiated. Given the societal context, the modernist fantasy will be permanently frustrated, because visitors' bodies and minds will not behave according to the preset requirements, and the dominance of cultural professionals fixating cultural meaning will be privately and publically contested and resisted by audience members, but also by other elites. Also within the cultural institutions themselves the modernist articulation of the cultural professional will be resisted, as is illustrated by Macdonald's (2001: 133) description:

> Those arguing for constructing the visitor as relatively ignorant were accused of being 'patronizing' and of 'dumbing down', those who constructed the visitor as more educated faced charges of 'elitism and of being potentially 'exclusionary'.

The modernist fantasy of the powerful and knowledgeable cultural professional also has aninverse variation, the democratic-populist fantasy which articulates the cultural professional as superfluous. In contrast to the othering processes which privilege the cultural professional, this democratic-populist fantasy is based on the replacement of a hierarchical difference by total equality. This fantasy remains embedded within a modernist framework because of its focus on equality. Moreover, it is a populist fantasy, because (following Laclau's approach) it is based on an antagonist resistance of the people against an elite. As Laclau (1977: 143) puts is:

> Populism starts at the point where popular-democratic elements are presented as an antagonistic option against the ideology of the dominant bloc.

This democratic-populist fantasy has a number of variations. The celebrative-utopian variation defines the equalization of society, and the disappearance of its elites, as the ultimate objective for the realization of a 'truly' democratic society. Cultural professionals become in this perspective problematized, as the symbolic power that is attributed to them is seen to be obstructing the process of cultural democratization. The process of equalization can be articulated as political, but also as economic, where the annihilation of hierarchical difference through capitalist market logics is met with approval. In this latter case, the notion of the cultural professional itself is transformed into a supplier of cultural goods, equalizing the power relationship between the suppliers and consumers of cultural goods. But there is also an anxietatic-dystopian variation, based on the fear that the democratic-populist fantasy might actually be realized. Here, the democratic-populist fantasy becomes supportive of the modernist fantasy of the powerful and knowledgeable cultural professional, as the democratic-populist fantasy serves at a constitutive outside for the modernist fantasy. One recent example is Keen's (2007) *The Cult of the Amateur,* where the 'amateurs' which produce user-generated content become seen as a threat to (expert) tastes, knowledges and truths.

Both fantasies remain firmly locked with a modernist framework, which renders them inherently problematic in the era of post/late/liquid modernity. The modernist articulation of the cultural professional (and the anxietatic-dystopian variation of the democratic-populist fantasy) privileges an elitist, Author-based model of society, where the construction of cultural meaning remains monopolized and the death-of-the-Author discourse is simply ignored. However resistant it is, this fantasy is in permanent conflict with the demands for opening up and democratizing the cultural field. The celebrative-utopian variation of the democratic-populist fantasy is equally problematic, because it conflates democratization with a stiffening equalization of society, and the reduction of power imbalances with the annihilation of difference. This radical denial of difference and its implicit rejection of "the best which has been thought and said in the world" (Arnold, 2004: 2), in other words of specialization, talent, expertise, and the Author, also finds itself in permanent conflict with society's structures and institutions, expert systems, discursive structures on cultural value and power dynamics. Moreover, both fantasies remain problematic because they are antagonistic, which is difficult to reconcile with the notion of democracy itself. To resolve this apparent deadlock, I want to turn to another fantasy, which I will term the participatory fantasy. Here, the starting point is Pateman's (1970) definition of 'partial' and 'full participation'. Partial participation is defined by Pateman as: "a process in which two or more parties influence each other in the making of decisions but the final power to decide rests with one party only" (Pateman, 1970: 70), while full participation is seen as "a process where each individual member of a decision-making body has equal power to determine the outcome of decisions" (ibid.: 71). The importance of Pateman's work is that it allows emphasizing the need for more balanced power relations in society (and not exclusively in the political system). Moreover, Pateman's definition does not imply that the position of (one of) the involved parties (in our case cultural professionals or audience members) should be erased. On the contrary, her definition entails a decision-making process that is respectful to all parties involved, on the basis of power sharing. This plea for an increase of societal power balances still has a clear utopian, fantasmatic dimension. Situations of full participation are utopian non-places – or, better, 'never-to-be-places' – which will always remain unattainable but which simultaneously remain to play a key role as ultimate anchoring points for democratization processes. Despite the impossibility to fully realize these situations in the social praxis, their fantasmatic realization serves as breeding ground for democratic renewal in the field of culture.

Simultaneously, we need to avoid the articulation of another modernist fantasy – this time when talking about participation – which ignores difference and the conflicts that difference brings about, or which frames differences as necessarily antagonistic. Here, we can turn to Mouffe's (2005) work, who suggested the concept of agonism to describe a "we/they relation where the conflicting parties, although acknowledging that there is no rational solution to their conflict, nevertheless recognize the legitimacy of their opponents" (Mouffe, 2005: 20). An agonist relationship does not hide the differences in position and interest between the involved parties; they are "in conflict" but "share a common symbolic space within which the conflict takes places" (ibid.). Translated to the participatory fantasy, this implies that the structural differences between cultural professionals and audience members are acknowledged, but that both parties accept that they share a common cultural space and accept each other's perspectives, however different they may be.

6. Conclusion

If we combine the agonism concept with the notion of (full) participation as outlined by Pateman, then we can describe this participatory fantasy as a respectful and balanced negotiation in cultural production processes, where all become authors (without a capital A) in interpretation and production, where difference is acknowledged, and where all voices can be heard and used to structurally (and not occasionally) feed the decision-making processes. This re-articulation of the cultural professional's subject position does not reject expertise, but recognizes different types of expertise. It does not reject public service, but sees the facilitation of participation as part of the public service remit. It does not reject ethics, but inscribes the equalization of power imbalances in the ethical framework of the cultural professional. It does not reject autonomy, but replaces one of its components, detachment, with connectedness. It does not reject institutional embeddedness, but respects amateurism (in Said's (1994) meaning[4]). The only identity component it does reject is the modernist privilege of the cultural professional to solipsistically detach him/herself from the social, without sharing his/her symbolic power.

Replacing fantasies is of course easier said than done, and the modernist fantasy of the powerful and knowledgeable cultural professional, and the equally modernist democratic-populist fantasy will not disappear. As fantasies, they remain important driving forces and sites of struggle that persist in the present-day cultural configuration. What the debates on the increase of participation in the world of cultural institutions shows is that people have managed to deconstruct these modernist fantasies and the subject positions that produce them (and are produced by them). We also see that this participatory fantasy was threatened by oblivion in the heydays of neo-liberalism and that only now the conditions of possibility of its resurgence have been created.

4 Said (1994: 84) defines amateurism as an activity that is fuelled by care and affection rather than by profit and selfish, narrow specialization.

One of the problems of participatory fantasies has been that they did not manage to cut themselves free from the modernist origins of the ideology of participation. Because of this modernist anchoring, it has been proven difficult to reconcile participation, difference and conflict, which unavoidably kept participation within the antagonistic framework of either the cultural professional as Author fantasy or its democratic-populist death of the Author counterpart. Possibly, the combination of participation and agonism might offer a much needed departure from this theoretical vacuum caused by an unnecessary dichotomization, allowing for the acceptance of difference in combination with an engagement towards more equalized power relations in processes of cultural production.

Acknowledgement

The publishing of this article was supported by the Estonian Science Foundation grant no 8006.

References

Arnold, Matthew (2004). *Culture and Anarchy*. Whitefish, MT: Kessinger.

Barthes, Roland (1984). *Image Music Text*. London: Flamingo.

Beck, Ulrich; Anthony Giddens, Scott Lash (eds.) (1994). *Reflexive Modernization: Politics, Tradition and Aesthetics in the Modern Social Order*. Stanford: Stanford University Press.

Bourdieu, Pierre (1984). *Distinction: A Social Critique of the Judgement of Taste*. London: Routledge.

Bourdieu, Pierre (2000). *Pascalian Meditations*. Stanford: Stanford University Press.

Bruns, Axel (2007). "Anyone Can Edit": Understanding the Produser – Guest Lecture at SUNY, Buffalo / New School, NYC / Brown University / Temple University. Available at http://snurb.info/index.php?q=node/286 (accessed June 25, 2007).

Cammaerts, Bart (2005). ICT-usage among Transnational Social Movements in the Networked Society: To Organise, to Mobilise and to Debate. Roger Silverstone (ed.) *Media, Technology and Everyday life in Europe. From Information to Communication*. Aldershot: Ashgate, pp. 53–72.

Carpentier, Nico (2005). Identity, Contingency and Rigidity. The (Counter-)hegemonic Constructions of the Identity of the Media Professional. *Journalism* 6 (2): 199–219.

Carpentier, Nico; Erik Spinoy (2008). From the Political to the Cultural. In Nico Carpentier, Erik Spinoy (eds.). *Discourse Theory and Cultural Analysis. Media, Arts and Literature*. Cresskill NJ: Hampton Press, pp. 1–26.

Castells, Manuel (2001). *The Internet Galaxy: Reflections on the Internet, Business, and Society*. Oxford: Oxford University Press.

Christians, Clifford G.; Kim B. Rotzoll, Mark B. Fackler (1991). *Media Ethics: Cases and Moral Reasoning*. New York: Longman.

Deuze, Mark (2005). What is Journalism? Professional Identity and Ideology of Journalists. *Journalism* 6 (4): 442–464.

Friedland, Lewis A. (1996). Electronic Democracy and the New Citizenship. *Media, Culture & Society* 18: 185–212.

Fuss, Diana (1989). *Essentially Speaking: Feminism, Nature & Difference*. London: Routledge.

Glynos, Jason, Yannis Stavrakakis (2004). Encounters of the Real Kind. Sussing out the Limits of Laclau's Embrace of Lacan. In Simon Critchley, Oliver Marchart (eds.). *Laclau: A Critical Reader*. London, New York: Routledge, pp. 201–216.

Hall, Stuart (1989). New Ethnicities. In Kobena Mercer (ed.). *Black Film, British Cinema*. London: BFI, pp. 27–31.

Halpin, Marjorie M. (1997). 'Play it Again, Sam': Reflections on a New Museology. *Museum International* 49 (2): 52–56.

Hemming, Steve (1997). Audience Participation: Working with Local People at the Geffrye Museum. In Eilean Hooper-Greenwill (ed.). *Cultural Diversity. Developing Museum Audience in Britain.* London: Leicester University Press, 168–182.

Hooper-Greenwill, Eilean (ed.) (1997). *Cultural Diversity. Developing Museum Audience in Britain.* London: Leicester University Press.

Huysen, Andreas (1986). *After the Great Divide: Modernism, Mass Culture, Postmodernism*. Bloomington: Indiana University Press.

Keen, Andrew (2007). *The Cult of the Amateur. How the Democratization of the Digital World is Assaulting Our Economy, Our Culture, and Our Values*. New York: Doubleday Currency.

Kirshner, Lewis (2005). Rethinking Desire: The *objet petit a* in Lacanian Theory. *Journal of the American Psychoanalytic Association* 53 (1): 83–102.

Krotz, Friedrich (2007). The Meta-process of 'Mediatization' as a Conceptual Frame. *Global Media and Communication* 3 (3): 256–260.

Lacan, Jacques (1979). *The Four Fundamental Concepts of Psycho-Analysis*. Ed. by Jacques-Alain Miller. London: Penguin.

Lacan, Jacques (1988). *The Seminar. Book II. The Ego in Freud's Theory and in the Technique of Psychoanalysis.* Notes by John Forrester. Cambridge: University Press.

Lacan, Jacques (1991). *Le Séminaire. Livre VIII: Le transfert.* Ed. by Jacques-Alain Miller. Paris: Seuil.

Lacan, Jacques (1994). *Le Séminaire. Livre IV: La relation d'objet*. Ed. by Jacques-Alain Miller. Paris: Seuil.

Laclau, Ernesto (1977). Towards a Theory of Populism. In Ernesto Laclau (ed.). *Politics and Ideology in Marxist Theory*. London: New Left Books, pp. 143–198.

Laclau, Ernesto (1990). *New Reflections on the Revolution of Our Time*. London: Verso.

Laclau, Ernesto (2005). *The Populist Reason*. London: Verso.

Laclau, Ernesto; Chantal Mouffe (1985). *Hegemony and Socialist Strategy: Towards a Radical Democratic Politics*. London: Verso.

Lash, Scott (1994). Expert-Systems or Situated Interpretation? Culture and Institutions in Disorganized Capitalism. In Ulrich Beck, Anthony Giddens, Scott Lash (eds.). *Reflexive Modernization: Politics, Tradition and Aesthetics in the Modern Social Order*. Stanford: Stanford University Press, pp. 198–215.

Lyotard, Jean-Francois (1984). *The Postmodern Condition. A Report on Knowledge*. Manchester: Manchester University Press.

Macdonald, Sharon (2001). Behind the Scenes at the Science Museum; London: Knowing, Making and Using. In Mary Bouquet (ed.). *Academic Anthropology and the Museum: Back to the Future*. New York: Berghahn Books, pp. 117–140.

Macdonald, Sharon (2002). *Behind the Scenes at the Science Museum*. Oxford and New York: Berg Publishers.

Manovich, Lev (2001). *The Language of New Media*. Cambridge: MIT Press.

Marstine, Janet (2006). Introduction. In Janet Marstine (ed.). *New Museum Theory and Practice. An Introduction*. Oxford: Blackwell, pp. 1–36.

McQuail, Denis (2008). Journalism as a Public Occupation: Alternative Images. In Nico Carpentier et al. (eds.). *Democracy, Journalism and Technology: New Developments in an Enlarged Europe. The Intellectual Work of the 2008 ECREA Media and Communication Doctoral Summer School*. Tartu: Tartu University Press, pp. 47–59.

Merriman, Nick (1989). Museum Visiting as a Cultural Phenomenon. In Peter Vergo (ed.). *The New Museology*. London: Reaktion Books, pp. 149–171.

Mouffe, Chantal (2005). *On the Political*. London: Routledge.

Pateman, Carole (1970). *Participation and Democratic Theory*. Cambridge: Cambridge University Press.

Penny, Simon (1995). Consumer Culture and the Technological Imperative. In Simon Penny (ed.). *Critical Issues in Electronic Media*. New York: State University of New York Press, pp. 47–73.

Reger, Jo; Daniel J. Myers, Rachel L. Einwohner (eds.) (2008). *Identity Work in Social Movements*. Minneapolis: University of Minnesota Press.

Rheingold, Howard (1993). *The Virtual Community: Homesteading on the Electronic Frontier*. Reading, MA: Addison Wesley.

Ross, Max (2004). Interpreting the new Museology. *Museum and Society* 2 (2): 84–103. http://www.le.ac.uk/ms/m&s/Issue%205/ross.pdf (accessed April 9, 2010).

Said, Edward W. (1994). *Representations of the Intellectual: The 1993 Reith Lectures*. New York: Pantheon.

Sandell, Richard (ed.) (2002). *Museums, Society, Inequality*. London: Routledge.

Sayyid, Bobby; Lilian Zac (1998). Political Analysis in a World without Foundations. In Elinor Scarbrough, Eric Tanenbaum (eds.). *Research Strategies in the Social Sciences*. Oxford: Oxford University Press, pp. 249–267.

Snow, David; Leon Anderson (1987). Identity Work among the Homeless: The Verbal Construction and Avowal of Personal Identities. *American Journal of Sociology* 92: 1336–1371.

Torfing, Jakob (1999). *New Theories of Discourse: Laclau, Mouffe and Žižek*. Oxford: Blackwell.

van de Donk, Wim; Brian D. Loader, Paul G. Nixon, Dieter Rucht (eds.) (2004). *Cyberprotest: New Media, Citizens, and Social Movements*. London: Routledge.

Vergo, Peter (ed.) (1989a). *The New Museology*. London: Reaktion Books.

Vergo, Peter (1989b). Introduction. In Peter Vergo (ed.). *The New Museology*. London: Reaktion Books, pp. 1–5.

Wilpert, Bernhard (1991). Property, Ownership, and Participation: On the Growing Contradictions between Legal and Psychological Concepts. In Raymond Russell, Veljko Rus (eds.). *International Handbook of Participation in Organizations (2). Ownership and Participation*. Oxford: Oxford University Press, pp. 149–164.

Wright, Philip (1989). The Quality of Visitors' Experiences in Art Museums. In Peter Vergo (ed.). *The New Museology*. London: Reaktion books, pp. 119–148.

Žižek, Slavoj (1990). Beyond Discourse-Analysis. In Ernesto Laclau (ed.). *New Reflections on the Revolution of Our Time*. London: Verso, pp. 249–260.

Žižek, Slavoj (1995). *The Sublime Object of Ideology*. London: Verso.

Identity Struggles of Museum Professionals: Autonomous Expertise and Audience Participation in Exhibition Production

Taavi Tatsi

1. Introduction

The following article is concerned with the formation of the museum professional's identity in two processes of exhibition production. One is a more traditional curatorial process and the other challenges such curatorship by opening up possibilities for structural audience participation in exhibition production. In order to analyse the 'traditional' formation of curatorial identity and what happens to it in processes where audiences are given more power, this article first looks at the identity processes in the ongoing production of the permanent exhibition on Estonian cultural history at the Estonian National Museum (ENM). Subsequently, this article juxtaposes this with an exhibition production process triggered by an audience empowerment project. The responses, such as resistance, anxiety and othering, make explicit both the consequences of the challenge to the established identity and also the limits to developing a more collaborative exhibition production model that are embedded in that identity. However, a more fluid/hybrid expert identity forms a perspective from which to theorise possibilities of overcoming some aspects in the dichotomy of experts and amateurs.

2. Methodology

This paper employs an ethnographic methodological framework. This comes through the author having taken part in the identity processes within the museum itself, both the more traditional curatorial processes and the challenge, the latter also including a facilitatory role in the project. Moreover, I have participated in these processes in a double role: first as a museum professional managing exhibitions, but after two years moving to the research department and becoming an ethnographer conducting participatory observation 'at home'. As a researcher, I had the advantage of already being immersed in the culture of exhibition production. In the permanent exhibition process, collecting data largely entailed taking part in the permanent exhibition planning meetings as a member of the curatorial team working on the exhibition content. From the processes of the intervention, the data is pooled through participatory observation at the intervention design meetings and the meetings where the project was introduced and discussed within a broader group of ENM professionals (open board meeting, internal seminar of the research department). Apart from that, I held a roundtable debriefing among the involved and interested museum staff after the first intervention exhibition had finished. Last but not least, the draft of the research article was circulated among the museum professionals in the ENM and feedback was encouraged.

3. Situating the theoretical issues and research object

Identities are here seen as social: possessing both individual and collective dimensions and working both towards establishing differences as well as similarities (Jenkins, 2008: 17–21). At the same time this article identifies with the fundamental social ontology of Ernesto Laclau and Chantal Mouffe (1985; also Laclau, 2007 [1996]) by believing that identities are contingent positions that identify with meanings. Meanings tend to be arranged according to some hegemonic discursive framework, but will also always preserve the possibility of being rearranged, identifying with other markers of meaning. The human knowledge of "who is who and what is what" is very much embedded in language and is a process (Jenkins, 2008: 5), therefore it can never be entirely fixed. Such processes are sometimes also called 'identity work' to capture "the discursive efforts that people have undertaken in order to (re)construct and maintain their identities" (Carpentier, 2011b: 189). The particular identity under study in this article is that of museum professional. It signifies those who are engaged in cultural production in museum institutions and employ curatorship over the museum collections and/or knowledge production, which involves constructing the frameworks of meaning enabled (or not) in museum exhibitions. There is one more important notion in the identity processes concerned by this research and that is participation. Without going into an extensive discussion of the term, 'participation' here is chosen to refer first and foremost to 'structural' participation which includes co-deciding exhibition content, policy and technology as well as evaluating the content (Carpentier, 2011a: 130). Whether it structurally incorporates audiences or only 'traditional' cultural experts is also a crucial aspect for analysing a traditional/established/modernist museum professional's identity when encountering a 'new' identity component that involves a different attitude towards a more structural audience participation. The museum itself, then, is the particular setting where the identity processes at the focus of this analysis take place. While traditionally "measured by its internal possessions such as collections, endowments, staff and facilities" (Watson, 2007: 1), it is becoming more and more influenced by "specific, demonstrable and measurable benefits to the public" (ibid.). In the discussions and debates over the social relevance of the museum (Fyfe, 2006) and the new horizons (often opposed to outdated limitations) for museums, referred to as "new museology" (Vergo, 1989) most emphasis is put on rethinking the museum's relationships with its audiences. While certainly not a completely new discourse (Dana, 1917), it has intensively brought to the forefront keywords such as 'access', 'social responsibility' and 'community involvement', replacing the discourses that emphasise collecting, interpreting and exhibiting (Witcomb, 2003: 59). The role of active audiences (Hein, 2006; Runnel, Pruulmann-Vengerfeldt, 2010) and the techniques through which to facilitate audience engagement (Black, 2005) and participation (Simon, 2010) as well as debates over empowerment and its limits (Macdonald, 2002; Golding, 2009; Burch, 2010) are increasingly at the heart of the museological texts, although they also in impact studies (Research Centre for Museums and Galleries, 2002) and policy documents (United Kingdom, Department for Culture, Media and Sport, 2005).

While Watson and Waterton (2010: 1) point out that community engagement/involvement as both concept and practice has been well integrated into the 'heritage sector', becoming part of the jurisdiction, legislation and management processes, sometimes even resulting in box-ticking and an ossification of the related assumptions and practices. However, the "abiding and inequitable imbalances between the professionals and communities" (ibid.: 2) might remain there, they argue. Relying on fieldwork in community-based archives, Stevens et al. (2010) show how archive professionals in the UK are not prone to valuing alternative forms of expertise, which clearly implies that the rigidity of the museum professional's identity is not only a matter for post-Communist societies. There is ample reason to invite museum (and heritage) professionals to reflect on the ways they construct and perpetuate certain components of professional identity and, through that, those of the community and audiences.

Informed by the issues raised in the overall intellectual climate of new museology, it is nevertheless important to take into account the Foucauldian genealogy of the museum and its agency in society theorised by Tony Bennett as a museum-specific "governmentality" the general regulatory aim of which is "to allow the people, addressed as subjects of knowledge rather than as objects of administration, to know; not to render the populace visible to power but to render power visible to the people and, at the same time, to represent to them that power as their own" (1995: 98). Another important perspective on the ways museums have functioned in cultural (often colonial) encounters is James Clifford's employment of Marie Louise Pratt's notion of "contact zone" by giving it a perspective for theorising encounters over social distances between the museum and communities "within the same state, region, or city – in the centers rather than the frontiers of nations and empires" (Clifford, 1997: 204). Andrea Witcomb does that effectively by analysing the pressures of popularisation on curatorial culture in a number of cases in the museum field of Australia.

She looks at the seemingly imperative but also very complex change in the curatorial culture from being centred on the traditional practices towards popularisation in two directions. The first of them is the "smiling professionalism" that marketing discourse calls for in order to survive economically in the increasingly competitive leisure market. This led, in one of the cases Witcomb studied, to the development of two different museum spaces: one based on the 'old' curatorial culture of thorough research and preservation expertise, and another one a leisure-market-oriented new space with a centrepiece that emphasised popular sentiments of the day in order to be more attractive to the general publics (Witcomb, 2003: 51–78). Witcomb also discusses community access galleries as another way of popularising museums, which may empower communities with the skills of curatorship and facilitate the production of their own representation(s) (ibid.: 79–101). This is another way of popularisation: implementing facilitatory practices (giving up power) but when encountering the curatorial culture of the museum, it also functioned in Australia as a community instruction project on cultural diversity. Here Witcomb acknowledges Bennett's arguments about the positive productive power of the museums' continuing governmentality and along similar lines questions the opposition between communities and museum that has been constructed by the new museology. Trying to overcome that opposition, Witcomb draws implications for the museum's changing role: by regulating communities, initiating civic reform and producing communities it is always possible to become more democratic and representative as new communities are continuously constructed and possible to reach. The curator and the museum cannot only play the role of facilitator but are destined to remain cultural producers as well (ibid.: 79–80). In other words, this can be viewed as an implication of a third way: museum professionals must become more reflexive and critical about their power(s), and become more open to diversified professional identities when it comes to traditional roles and functions of a museum. At the same time, they continue to work from the position claiming (productive) power and (positive) governmentality inherited from the modernist agenda of museum professionalism.

As Tali and Pierantoni point out, the construction of new museum institutions in Central and Eastern Europe tend to be driven much less by the local civic society than the interests of public authorities and neoliberal market actors. These new institutions all consequently tend to function as symbolic monuments for the new social order established since the beginning of the 1990s (Tali, Pierantoni, 2008: 243, 259–260). After the Estonian Academy of Music and Theatre and the Art Museum of Estonia (Kumu), the ENM has been third on the official construction waiting list of such symbolic monuments, being also an old debt to the national consciousness by not having had its own building, created for museum purposes, throughout the 103 years of its history.[1] At the time of writing, the general building process has passed through the phases of international architectural competition and preliminary design. The production of a permanent exhibition is in the middle of conceptualisation, design and object selection. The current permanent exhibition from 1994, on Estonian culture, is essentially a display of an ethnographic present of late 19th century peasant culture and is based on the collecting ideology that once sparked the establishment of the ENM. The prospective display, however, aims at broadening the scope both chronologically and paradigmatically by extending the beginning of the storyline from earliest history to the present day and discursively also looking for a more multicultural and diverse representation of everyday life (Rattus, 2009).

4. Identity formation of the museum professional in 'traditional' exhibition production

Those involved in the permanent exhibition production are ENM professionals. They form the core curatorial team, although there are several external experts involved too, from the fields of archaeology, language studies, folklore studies and religion sociology. The architects of the building and exhibition designers are also external experts, commissioned to the design task through a competition. The external experts assume the position of a museum professional, but there are also effective distinctions at work to differentiate between the museum's own expertise and other experts in the process.

Following the theoretical framework of social identity in Laclau and Mouffe's discourse theory, Nico Carpentier (2011b) formulates the "old"/modernist components of the cultural expert's identity. The first of them is knowledge and skill, expertise in context and objects, sometimes with a more contemporary marketing and managerial knowledge component (legitimate knowledge in Bourdieuan terminology). Closely linked to expertise is the second element – autonomy from a number of influences, such as the market and state but in some situations also audiences with their "unwarranted claims". The third element is public service provision, which tends to (but not necessarily) articulate audiences as more passive receivers. A certain professional ethic forms the fourth element, to which non-experts are not bound. The fifth element structuring the culture professional's identity is institutional embeddedness, which is often based on employment relationship, support systems and a network of peers. The formation of a professional community can lead to the positioning of audiences as 'ordinary' thus making it distinct from pro-

1 Estonian National Museum was established in 1909 by leading intellectuals of the national awakening, nine years before an independent country of Estonia appeared on the political map.

fessional 'elites'. Stemming from the professional responsibilities, a cultural expert inevitably deploys management and power, and this sixth element leads to a sense of psychological property (Carpentier, 2011b: 191–193). Consequently, cultural production often entails the "management of audiences' bodies and the targeted exposure of audiences' minds to carefully selected meanings" (ibid.: 193).

In the ENM exhibition production, museum curator identity is embedded in the habitual practices and articulations. On closer examination, its formation resembles first and foremost an established/traditional/modernist identity of a culture professional, with some institutional diversity challenging this hegemonic discursive structure. At the time of writing, the identity work of museum professionals and the involved experts goes on between themselves, retaining autonomy from the public. These culture professionals thus 'govern' the visitor experiences and construct the museum exhibition space. While doing so they maintain and generate the power that enables them to construct 'appropriate' narratives and object displays with the 'appropriate' tone and design. Even though it is possible to conclude that the museum professional can rely on the bastions of its autonomy, it has not been an easy process to synchronise curatorial knowledge and the knowledge of professionals from the other fields of production – especially with those of the architects and the designers. Extensive struggles between curators and architects over where to position the internal walls are quite exemplary here. While the debates were hardly over the meanings that the architecture offers they reveal how the autonomy of the culture professional, even when well defended from the influence of the audiences, can also be a source of antagonism when different types of legitimate knowledge claim authorship over the same area. Curators have been engaged in similar battles with the designers too as in the process of prototyping the curators have sometimes experienced that some design choices override their authorship over content. Such struggles are controversial and uncomfortable, but nevertheless appear at the same time to be legitimate compared to the hypothetical ones that structural audience participation would give rise to. There is a default agreement that those who are engaged are entitled to the position of cultural expert and thereby to deploy their legitimate power over structure and content, which overlap to some extent and (because knowledge and professional ethics might differ) create antagonisms/struggles.

The struggles with the external experts simultaneously appear to prevent the internal antagonisms within the core curatorial team from appearing, which could theoretically stem from different positions regarding the pressures of popularisation. In the case of the ENM, there has not been as much pressure to move towards marketing-oriented popularisation as in the Australian cases that Andrea Witcomb has analysed. The traditional functions of the museum are still dominated by the structure of its departments: collections, research, conservation, exhibitions and museum education. The relatively high autonomy of the curators from the influence of the market can to some extent be explained by the fact that there is no powerful marketing unit at the ENM as of today and that the managerial and marketing component of the museum professional is not widely incorporated within the curator identity. The leading role in permanent exhibition production is designated to the curators of the research department, many of whom have experience of producing exhibition content over the years in the temporary building. Their 'legitimate knowledge' largely stems from ethnography (historically centred on material culture) and cultural history. A small minority are also members of the intervention design team, embedded in (new) media and communication studies, also influenced by new museology. The head curator of the permanent exhibition is also clearly informed in contemporary anthropological theory and cultural studies, leading to the advocacy of multiculturalism and detachment from the ways of reconstructing an ethnographic present employed in the current permanent exhibition on Estonian cultural history. The common ground for curating content is currently broadly defined as 'everyday life' where the sources of data are 'informants'. The researcher-informant relationship could be theorised as a certain way of facilitating audience participation (what Nina Simon (2010) categorises as contributive participation) and through that giving legitimacy to exhibition producers to do their work. Although the representation of different cultural (mostly ethnic) minorities has been on the agenda of the permanent exhibition, the more structural participation of audiences has been incorporated in a very limited way in what seems by default be a full-scale professional game.

The museum professional here clearly has to work in partnership with the external parties, for whom the museum professionals represent the client and, paradoxically but unsurprisingly, sometimes even the public (Ghotmeh, 2009). The implicit premise seems to be that the public will by default benefit most from the end product from the best "public serving" experts.[2] The identity components of autonomy and public service seem to hide an antagonism: while the experts work autonomously to the greatest public good, there is a tendency to underrepresent the public because the museum professionals themselves would be in the (power) position of a client in that process. Consequently, the audiences become slightly annihilated symbolically, resembling the visualisations of the designers where semi-transparent human figures stroll through the exhibitions sometimes engaging in predesigned interactions.

2 See Runnel et al., 2010 for a more thorough discussion about the divide between professional expertise and the lack of dialogue with the public when attempting to 'reinvent' the Estonian National Museum through the design of a new building.

5. Challenges to the museum professional's identity in the Open Curatorship project

In parallel to the permanent exhibition production process, an intellectual intervention was designed, stemming from a broader research project agenda influenced by new museology, cultural studies, and new media studies. The design came from a research group affiliated both with the museum research department and the university. The research project design is built on a number of interventions that follow an academic agenda, but are at a more practical level also aimed at introducing some new communicative approaches to the everyday work and practices of the ENM. The research group has, in two and a half years designed five different audience participation interventions at the museum and led internal seminars on museum communication (focusing on new media and audience participation). In the context of this article these are interpreted as diversification opportunities for rearticulating the established identity components of museum professionals. The intervention of particular interest here is the Open Curatorship model – publicly promoted as "Create Your Own Exhibition" – as the one aiming to develop structural audience participation in the field of exhibition production. What took place was, in short, the launch of an open public call inviting everyone except museum professionals to submit their ideas for an exhibition in the ENM temporary exhibition space, with a public online/onsite vote conducted to determine the winners.[3] It has been possible to submit exhibition ideas to the ENM in the past, but it has not been strategically communicated to the general public before and a committee of ENM professionals has always been the sole gatekeeper making the decisions on who gets to make an exhibition at the museum.

In the framework of the intervention, the power relations between curators and audiences are played out differently, and this consequently presents an obvious challenge to the established identity of the museum professional by restructuring roles and redistributing power. On the one hand, the museum professional in the Open Curatorship production format can, instead of fully controlling exhibition content and design, set minimal terms and conditions to the process where publicly selected members of the audience make decisions over museum content. On the other hand, the museum professional's identity was provided with an opportunity

3 There were a total of 33 proposals for the "Create Your Own Exhibition" project (27 applicants with their own objects and 7 engaging museum objects) and 564 voters participated online and onsite to choose the two winners: one with the applicant's own exhibits and the second that engaged museum collections as well. The two proposals that won the contest went into the exhibition production process and involved museum staff from exhibitions manager to public relations person, as well as collection managers and conservators.

to embrace new components. In order to analyse the diversity of the responses in the museum context, this paper brings in a theoretical framework of these possible components from a comprehensive analysis by Carpentier (2011b) in an analysis of a culture professional's identity in general. These are modelled after what he calls "an agonistic participatory fantasy" stemming from the socio-cultural actuality of a "more post/late/liquid-modernist logic" (ibid.: 2). According to that model (which is more or less echoed in the overall agenda of the intervention):

> (1) The knowledge and skills component established by legitimate knowledge could be 'updated' with recognition of the diversity of expertise during the employment of curatorial skills in the national museum exhibition space.
>
> (2) The autonomy component would have to avoid detachment and anxiety towards audience participation and employ well-communicated connectedness instead.
>
> (3) Public service provision could entail more facilitation of participation.
>
> (4) The fundamental professional ethics should accordingly encompass the principle of equalising power imbalances in the skills and resources needed for exhibition production and, while remaining embedded in institutions and peer networks, foster a respect towards amateurism by finding new ways to include amateurs in these networks and even institutions.
>
> (5) Last but not least, while continuing to deploy management and power over the museum collections, museum professional would have to explicitly communicate how such a symbolic power could be shared in an exhibition space. (Carpentier, 2011b: 200–201)

When looking at the responses to the offered new components, it is apparent that these rely strongly on the established discursive structure of a museum professional's identity, and a hope to see a quick assimilation of a participatory identity work seems to be rather idealistic. The responses to the challenge were pooled at three instances during the different phases of the project. The first was before the project had actually been announced in public – an open board meeting of the ENM (a regular practice, discussing issues approximately once a month in the organisation with the wider forum of the museum staff) where the overall activity of the research group was introduced and the "Create Your Own Exhibition" project proposal was intensively debated as a fresh project idea. The second meeting providing responses was one of the internal museum communication seminar series where the soon to be launched project was thoroughly debated. At that time, the project had just begun and a few initial ideas had already been submitted. The third meeting took place as a debriefing session after the first winning exhibition had just been taken down from display. Not as large a number of museum professionals participated, but the discussion was more focused and responses more diverse (two of the post-production responses came by e-mail). The first two debates involved more museum professionals and the challenge was much more 'imaginary' than at the debriefing session. In addition, the first two meetings evoked proportionately much more resistance, anxiety, othering and (with only a couple of exceptions) no supportive assimilation. The diversity of the responses was largely only in the different articulations of resistance/othering and also in the level of anxiety. It was only after the first production process had come to an end that the responses diversified.

Positioning from a potential selection committee member to becoming an voter equal in the selection process with 'ordinary' people does create a threat to the established hegemonic professional identity. Resistance was expressed against the way that the exhibition to be produced was chosen, i.e. to the new scheme of power relations, which created a feeling of museum professionals being left out of the decision-making process. The traditionally receptive/passive audiences were being 'upgraded' to a position of an active content-provider and decision-maker, which the established professional identity began to resist also by doubting whether the members of the general public could really refrain from exploiting the possibility of voting more than once and for oneself. The responses at the start of the project also revealed anxiety echoing through the resistance over whether the "Create Your Own Exhibition" project would create a conflict over museum resources by, for example, claiming the same exhibition space at the same time as when the museum would want to use it. Such attitudes show that the museum professional is perhaps too comfortable with the professional committee making decisions on museum exhibition content and programming, and leaving that to the audience is difficult to integrate within the established elements of their identity. The potential amount of extra work it could cause created similar anxiety over the possibility of overexploiting museum resources (with a project that might not be the museum's top priority) and the key role of professional skills regarding the object management in exhibition production were highlighted.

One of the aims of instituting the category of engaging museum objects into the "Create Your Own Exhibition" format was to intervene in the established ways of contributory public-collections relationship. These are traditionally based on a correspondents' network established in 1931, gathering mainly textual answers to thorough questionnaires, but also photographs and objects. Access to the actual objects once they are in the collections is highly regulated and a member of the public accessing these objects is positioned with the signifier 'researcher', which seems to imply the privilege of 'serious' research interests towards the 'authentic' objects (such as a museum professional would have) over all other kinds of interest or participation. Museum collections are at present managed by different museum professionals and, for a number of infrastructural and historical reasons, are geographically rather dispersed into spaces not directly accessible to the public. This hasn't left much room for even considering someone other than a museum professional or acknowledged cultural expert participating in making an exhibition which engages museum objects. In order to overcome the rigid dichotomy of cultural expertise and amateurism, the intervention experimented with the possibilities of opening up forms of 'third expertise' to enhance access and participation related to museum collections. There was a clearly communicated opportunity to propose exhibition ideas with museum objects in the Open Curatorship intervention. The possession of legitimate knowledge provides the traditional museum professional's identity with the power position required to doubt whether giving audiences the power over which objects to show from the museum collections is the right thing to do: "people, even our younger [colleagues] do not know the collections" and whether "is it really possible to present an idea with a picture of something (s)he has not even seen?" (author's fieldwork notes). The established premise is to keep ignorance at bay and 'educate' the public through professionally curated exhibitions rather than seeking participatory opportunities to overcome that ignorance in alternative ways.

Another type of response was to other the participating audiences by articulating them as authentic when they were as autonomous from professional influences as possible. This can be characterised as a very professional-centred point of view because they appear to prefer to be autonomous from the influence of the amateur audiences. Ways of constructing the true amateur identity in the exhibition context were consequently opposed to culture professionals at general whose ideas might also not be suitable for such an exhibition format. Even though these other cultural experts are not museum professionals, their knowledge appears not to be 'legitimate' in this context. Potential participants were signified mainly with an idealised non-professionality, whose value to the museum depends on how well their 'amateurishness' becomes evident through the exhibition. This is again a museum-professional centred way of looking at the possible identities in the exhibition production process.

However, the possibility of the 'third expertise' is constantly undermined by these antagonistic significations in opposition to a 'real' museum professional. Both the intervention designers and the other museum professionals shared the anxiety that exhibition proposals would be dominated by amateur collectors (e.g. matchboxes), hobbyists (e.g. painters) or performance artists. Such audiences were suspected of being incapable of sticking to the right topic (everyday life) because of too little reflection over how what they want to exhibit expresses the everyday. One of the proposals given as a negative example here was an idea to exhibit someone's nature photos, which raised the issue that it is not an expression of everyday life because an explicit human element was missing from the frame. In other words, there are some human expressions less desired by museum professionals to qualify as an exhibition displayed at the ENM: where the established professional identity prefers traditional curatorship over the possibility of participation-sensitive 'third expertise', there the interventionist seeks to connect audiences who have not yet discovered the legitimacy of their knowledge and experiences in the museum exhibition context. Both share the assumption that those who want to participate are more likely to have an almost abnormal drive to exhibit but are less likely to have the appropriate content for a national museum exhibition space. Here it is also important to note that the agenda of the intervention was also intended to reach out to audiences who would otherwise not imagine a (national) museum being relevant for their lives in any way; the agenda also influenced the ways the potential participants were imagined.

Regarding the museum collections and possible 'third expertise', the debate raised by those representing the established professional identity was not actually about how to raise and facilitate public interest in collections, but much more about finding arguments for how to defend against an imagined rush by the public into the collections in preparation for an exhibition idea. This is reflected in a statement by a museum professional that excursions to the collections are definitely out of the question, thus also discarded as a potential strategy to overcome the obstacle of lacking legitimate knowledge about the collections. There was a general feeling one could sense that the regular access hours and online directories for getting acquainted with collections were somehow not enough in the context of the Open Curatorship project, although in actuality it proved to be more than enough. All these can be interpreted as indications of building (traditional) identity bastions related to decision-making, legitimate knowledge and collections. The museum serves the public, but at the same time there seems to be a tendency towards a stiffening of identity regarding this service especially when attention is drawn to alternative access and participation approaches for the audiences consulting the collections. While collection managers are seen as a valuable resource, seeing 'third expertise' along similar lines is not yet a reality.

This introduces the argument surrounding another important established component of museum professional identity that is perhaps most difficult to attribute any kind of 'third expertise', namely professional ethics. The lack of it is echoed through an experienced danger that the audiences would act irresponsibly when producing the exhibitions: because the process is too complicated, the person(s) might not be ready to comply with all the proposals given and terms/conditions set by the museum. The most dubious expression of othering through the lack of professional ethics was a spreading rumour in the museum (when the first "Create Your Own Exhibition" was already on display) that a few old photographs put on display by the 'amateur curator' had been stolen from an old house in the countryside. This was never confirmed, but it left an overall subconscious suspicion towards the diversification of expertise in exhibition content production.

After the first exhibition production the responses were more diverse and slightly more reflexive, although the traditional identity remained well established and the consequences for the professional identity become more evident. After having seen the first "Create Your Own Exhibition",[4] some of the museum professionals involved in the production struggled to articulate the possible benefits of the project for the museum, while doing so still largely giving their opinions based on the standards of (modernist) curatorship. Those more involved in the actual production experienced that the "Create Your Own Exhibition" curator was working very independently; the interventionist facilitator helped with the logistics so that workload was not particularly high and regarded positively as such. The opportunity to engage 'other kinds' of audiences was also generally seen as positive, but when it became evident that they might not always synchronise with the museum professional's established view on what an exhibition should be like, some antagonisms become apparent, leading to defensive identity work. The first of such identity bastions is the value of objects from the perspective of what would contribute to the existing course of exhibitions at the museum as well as to the museum collections. The general conclusion was that the first own exhibition, apart from being emotionally difficult for some because of its topic, did not provide any new paradigm or approach. The bottom line for evaluating the success for the museum was not the participatory characteristics of the production process but whether objects on display were already featured in the collections and whether the exhibition (or a submitted idea) wasn't too focused on a myriad of objects, sometimes coupled with a similarly undesirable overly historical perspective. Here, again, an authentic amateur was constructed as ideally bringing "new quality in content and design", which once again evokes the antagonism of amateurs being authentic because they are not professional but simultaneously becoming othered as such.

What clearly emerged after the participatory intervention was the established museum professional identity working towards a clear distinction between the museum exhibitions and the public's own exhibitions, sometimes desiring this to be explicitly reflected in the design. A professional involved argued that probably no

4 The first "Create Your Own Exhibition" that also won the idea contest was on Estonian funeral traditions and customs, interpreted by a funeral director who collects related materials and objects, and has written a self-published book on the topic.

more than one out of five visitors was aware of the fact that this was a public and not an ENM exhibition. Another proposed trying to delegate even more power to the participating audiences and aim at a fully hosted[5] exhibition format in the future. However, such a separation (even when it is articulated as a productive 'branding' project) in the museum between professional and amateur exhibitions, might well lead to the creation of a relatively isolated culture of hosted exhibitions that would not function as a contact zone between the museum, participants and audiences. Neither would it facilitate the productive governmentality of the museum very much, because such a model would not empower audiences with the curatorial knowledge and skills that they are constantly argued to be lacking. This is not to argue that hosted exhibitions do not have place in a national museum – on the contrary, they are already taking place regularly and the Open Curatorship format might want to aim at more collaborative exhibition production, which requires new participatory components to be integrated into the hegemonic museum professional's identity.

6. Conclusions

The Estonian National Museum has not yet had the chance to produce a truly contemporary display on Estonian culture in a true museum building. This is a debt to museum culture to be paid off. It coincides, however, with the sociocultural changes that put the museum's relationship with its audiences under review, meaning that there is a myriad of changes to be implemented at the same time.

The third way implied in Witcomb's analysis on the pressures and changes in curatorial culture and Carpentier's model of a more participatory identity for the culture professional does remain both a chance and a challenge in the Estonian case. The ENM professional has so far been able to enjoy relatively low pressure from marketing-oriented popularisation and has at the same time retained and defended the 'old' modernist identity of museum professional. From the perspective of that identity, the culture of producing the new permanent exhibition is centred on facilitating traditional/established professionalism of the expertise related to the field, engaging different professionals who participate by applying the best practices of their fields. At the time of exhibition production today, structural consultations take place between experts, (re)interpreting the existing collections and filling in the gaps according to the needs of the constructed abstract narratives while imagining a community of visitors. The potential in the developments of new media are high on the agenda of the designers and emphasise both access and

---

5 In her book, *The Participatory Museum*, Nina Simon (2010) distinguishes four main types of participatory projects: contributive, collaborative, co-development and hosted. Her terminology echoes in the discussions of this article.

interactivity. Communities of today are largely left with the opportunity to consult a readymade exhibition when it is opened. Structural participation is looking overall to be quite limited, but the open access gallery will probably be developed in the climate of participatory design. Runnel et al. (2010) have argued that there is, however hardly any consultation, not to speak of audience participation, regarding the permanent exhibition spaces. Although it is never clear whether and to what extent audiences are ready to take some of the responsibilities usually 'delegated' to the cultural expert, such a structure and the invitation to it has to come from within the museum both at the rhetorical (already appearing from time to time) and practical level, with different modes for participation gradually integrated to the permanent exhibition.

When it comes to the Open Curatorship intervention, the critical arguments of museum professionals construct their own identity by positioning themselves against alterity (i.e. the audiences) by signifying them with what a professional is not. What flashes in these discussions is the museum professional not (yet) willing to symbolically share the stage of museum exhibition production with amateurs by employing more diverse, hybrid and negotiated participatory identities and doing that on more equal and empowering terms. A significant obstacle is anxiety about the museum professionals' own acquired and established professional standards (and with that, their established identity) being damaged or watered down. Keeping in mind one of the important components of the modernist culture professional's identity – deployment of power –, then at the heart of the intervention is a relatively strong disempowerment of the museum professional and an empowerment of the audiences by asking them to provide content and participate in voting to determine the winners. Both components of the intervention were unprecedented as such at the ENM. The terms and conditions were set so that the museum professionals were not allowed to participate in idea submission and the vote was also in stark contrast with the traditional process in the ENM where the exhibition programme is decided in a committee comprised of relevant museum professionals. What the intervention offered was a new facilitatory and participatory identity, reconfiguring the museum professionals' position to that of being a partner for the empowered audience, instead of an autonomous decision-making body of who gets to see what and which meanings are available at the museum to the general public. There was a significant amount of resistance, which implied a clear-cut distinction between museum professionalism and the professionalism of the Open Curatorship model, in which 'third expertise' is given more control. What the Open Curatorship format seems to be facing in the museum setting is a need to develop a way to very clearly communicate the fact that exhibitions can be produced in a climate of a 'third museum'. This communication would probably have to articulate possible rearrangements of the identities of researcher and informant in the way knowledge is produced. Apart from that, it is at the same time crucial to give participating audiences the opportunity to become signified as respected 'third experts': audience-as-curators of their own content. This calls for the integration of the autonomous curatorship skills and knowledge of museum professionals relating to exhibition production into a more collaborative (and inevitably agonistic) public agora for proposing and producing museum exhibition content. The museum professional needs therefore to be assured that audience participation does not make things too complicated for them. Those engaged in the relevant identity work need to co-produce and acknowledge the benefits for the museum that have so far been developed in the spirit of (high) modernism into a more democratised cultural sphere with a newly legitimate sense of a shared responsibility and symbolic space. Such a professional identity in a museum experiments with the construction of a participatory climate in order to be able to integrate the collaborative/participatory component as a valuable and necessary component of a museum professional's identity.

What takes place in a museum is a production of a particular culture of knowledge filtering, layering, design and display. When there is integration of the connectedness and sharing symbolic space in museum professionals' identity, a constant context-dependent negotiation over this contact zone of particular cultures on more equal terms could gradually come into being as an acknowledged exhibition format at the ENM. It would become a sort of 'third museum' where the museum professional doesn't only administer, but actively climbs on the ladder of participation together with the audience-as-curator to exhibit and contextualise content according to the particular negotiated agenda, relying on collaborative knowledge and skills. Through that, the ENM would increasingly act as an agent reforming the public (Bennett's governmentality in the positive sense) towards a civil society, provided that what are produced there are not only comfortable truisms but also diverse contact zones between the cultures of audiences and museum culture, thus also producing new cultures.

One obviously cannot expect a new national museum to be composed of halls full of participation and community access galleries. The (high) modernist museum agenda of established, but communicatively uni-directional displays of cultural content will always be there. Hopefully, the new permanent exhibition area of the ENM might benefit from operating not only access or even interactional types of participation but also more structural ones. And the uses with which the visitors, users, and audiences will be engaging themselves might provide valuable input from which to set new landmarks of participation in civic society as a whole – definitely a desirable strategic goal for the ENM in the 21st century.

Acknowledgements

The publishing of this article was supported by the Estonian Science Foundation grant no 8006.

Sources

Author's fieldwork notes, April 2008 – March 2011.

References

Bennett, Tony (1995). *The Birth of the Museum: History, Theory, Politics.* London: Routledge.

Bennett, Tony (2006). Civic Seeing: Museums and the Organization of Vision. In Sharon Macdonald (ed.). *A Companion to Museum Studies.* Malden, MA; Oxford; Carlton: Blackwell, pp. 263–281.

Black, Graham (2005). *The Engaging Museum. Developing Museums for Visitor Involvement.* London; New York: Routledge.

Burch, Stuart (2010). Taking Part: Perfromance, Participation and National Art Museums. In Simon Knell et al. (eds.). *National Museums. New Studies from around the World.* London; New York: Routledge, pp. 225–246.

Carpentier, Nico (2011a). *Media and Participation. A Site of Ideological-Democratic Struggle.* Bristol; Chicago, IL: Intellect.

Carpentier, Nico (2011b). Encarando a morte do Autor. O trabalho identitário do profissional de cultura e as fantasias de controle. Facing the Death of the Author. Cultural Professional's Identity Work and the Fantasies of Control. *MATRIZes*. Vol. 4, No. 2: 183–204.

Clifford, James (1997). *Routes: Travel and Translation in the Late Twentieth Century.* Cambridge, MA; London: Harvard University Press.

Clifford, James (2004). Looking Several Ways. Anthropology and Native Heritage in Alaska. *Current Anthropology.* Vol. 45, No. 1: 5–30.

Dana, John Cotton (1917). *The Gloom of the Museum.* No. 2 of the New Museum Series. Woodstock, VT: The Elm Tree Press.

Fyfe, Gordon (2006). Sociology and the Social Aspects of Museums. In Sharon Macdonald (ed.). *A Companion to Museum Studies.* Malden, MA; Oxford; Carlton: Blackwell, pp. 33–49.

Ghotmeh, Lina (2009). Designing an "Open" National Space. Paper presented at the Estonian National Museum 50th annual conference *National Museums as Spaces of Cultural Diversity and Dialogue*, Tartu, 13 April.

Golding, Viv (2009). *Learning at the Museum Frontiers: Identity, Race and Power.* Farnham; Burlington, VT: Ashgate.

Hein, Hilde S. (2006). *Public Art: Thinking Museums Differently.* Lanham, MD; Oxford: Rowman & Littlefield Publishers.

Jenkins, Richard (2008). *Social Identity.* (3rd ed.) London; New York: Routledge.

Laclau, Ernesto (2007) [1996]. *Emancipation(s)*. London; New York: Verso.

Laclau, Ernesto; Chantal Mouffe (1985). *Hegemony and Socialist Strategy: Towards a Radical Democratic Politics*. London; New York: Verso.

Macdonald, Sharon (ed.) (2006). *A Companion to Museum Studies.* Malden, MA; Oxford; Carlton: Blackwell.

Research Centre for Museums and Galleries (2002). A Catalyst for Change. The Social Impact of the Open Museum. Leicester: RCMG. Available at http://www.le.ac.uk/ms/research/Reports/catalyst.pdf (accessed April 26, 2011).

Rattus, Kristel (2009). From the Museum of Peasant Culture towards the Museum of Cultures. Paper presented at the Estonian National Museum 50th annual conference "National Museums as Spaces of Cultural Diversity and Dialogue", Tartu, 13 April.

Runnel, Pille; Pille Pruulmann-Vengerfeldt (2010). Muuseumikommunikatsioon ja kultuuriline osalus Eesti Rahva Muuseumis. [Museum communication and cultural participation at the ENM.] In Pille Runnel, Tuuli Kaalep, Toivo Sikka (eds.). *Eesti Rahva Muuseumi aastaraamat* 53. [A Yearbook of the ENM. No. 53] Tartu: Eesti Rahva Muuseum, pp. 118–132.

Runnel, Pille; Taavi Tatsi, Pille Pruulmann-Vengerfeldt (2010). Who Authors the Nation? The Debate Surrounding the Building of the New Estonian National Museum. In Simon Knell et al. (eds.). *National Museums. New Studies from around the World.* London; New York: Routledge, pp. 325–338.

Simon, Nina (2010). *The Participatory Museum.* Santa Cruz: Museum 2.0.

Stevens, Mary; Andrew Flinn, Elizabeth Shepherd (2010). New Frameworks for Community Engagement in the Archive Sector: From Handing over to Handing on. *International Journal of Heritage Studies*. Vol. 16, Nos. 1–2: 59–76.

Tali, Margaret; Laura Pierantoni (2008). The Processes of Contemporary Museum Constructions: Designing Public Space and Engaging Audiences. In Peter Aronsson, Andreas Nyblom (eds.). *Comparing: National Museums, Territories, Nation-Building and Change, NaMu IV, Linköping University, Norrköping, Sweden, 18–20 February 2008: Conference Proceedings.* Linköping: Linköping University Electronic Press.

United Kingdom, Department for Culture, Media and Sport (2005). *Understanding the Future: Museums and 21st Century Life – The Value of Museums.* Available at http://www.culture.gov.uk/images/consultations/UnderstandingtheFuture.pdf (accessed April 26, 2011).

Vergo, Peter (ed.) (1989). *The New Museology*. London: Reaktion Books.

Watson, Sheila (ed.) (2007). *Museums and their Communities.* London; New York: Routledge.

Watson, Steve; Emma Waterton (2010). Heritage and Community Engagement. – *International Journal of Heritage Studies*. Vol. 16, Nos. 1–2: 1–3.

Witcomb, Andrea (2003). *Re-imagining the Museum: Beyond the Mausoleum.* London; New York: Routledge.

Democratising Collections through Audience Participation: Opportunities and Obstacles

Taavi Tatsi

Agnes Aljas

1. Introduction

Collections of objects have historically been the most important element distinguishing museums from other cultural institutions. Today we can find museums without collections, some whose collections are only digital, and those focusing their activities mainly on exhibitions, community involvement and education. The guiding research question of the analysis below is: how has participatory democratisation influenced museum collections? And, more specifically, what opportunities and obstacles become evident in these processes where legitimising new and existing objects for the museum collections are concerned?

This paper will discuss the interactions between the museum's structure of legitimising objects and audience participation processes at the Estonian National Museum (ENM). This museum is undergoing a number of significant changes, such as the construction of a new building, a process of renewal that could bring about a museological shift, among many other things. In order to study and contribute to the development of museum communication, a group of researchers – including the authors of this article – designed a sequence of research interventions. The interventions are part of a research project that has two major (interacting) objectives: first to study the relationship between the ENM and its audiences (in general); and secondly to study the possible conditions and development of museum communication and audience participation through a series of (onsite and online) interventions. This paper analyses 4 of these interventions, carried out from 2009–2011 (2 interventions are left out of the data set because one of them had not finished by the time of writing this paper and the other primarily addressed the interaction between audience and display in the exhibition space). The interventions are related to two on-going participatory processes at the museum.

The empirical material for our analysis comes from the auto-ethnography and production ethnography of these interventions.

2. Collections and audience participation in the contemporary museum

In her overview of the role new museology has played in Australia since the 1980s, Claire Baddeley points out that despite the attempts to democratise society by becoming increasingly active agents (greater collaboration with community groups, facilitation of interpretive multivocality and sensitivity towards social diversity), "a number of museums in Australia continue to wrestle with the value, integrity and importance of collections and how to best display them" (2009: 92). The pressure towards becoming more visitor-centred obviously brings about debates that are in some cases tendentious in nature (Appleton, 2007), as well as the more balanced reflections (Knell, 2007) that take into account intellectual and social trends while at the same time being careful not to devalue the professional expertise developed through the care of museum objects. Another important aspect of this issue is discussed by Srinivasan et al. (2009): "what implications have the 'promises' of new museology had for museums along with the consequences of the advent of web 2.0 technologies".

The inclusion of diverse audiences has, according to their study, often been "largely temporary, limited to the life cycle of the rotating exhibition or event and outside of the museum's information record – it's catalog" (ibid.: 265). Their focus is on the destandardisation of the museum catalogue and the ways in which inclusion and participation, mediated by digitised collections, can engage audiences and extend the definition of expertise in the museum context. Thus, Srinivasan et al. recognise the importance of objects and advocate structural changes in museum collections enabled by the online and digital participation opportunities. It is important to note that the aim of this analysis is by no means to devalue the work of professionals, who accession, preserve and study objects in museum collections, but rather to find a balance between the contemporary currents of democratisation and acquired professional expertise as it relates to museum collections. When participatory interventions encounter the existing professional practices at the ENM, it is also crucial to elaborate the concept of participation in a museum context, where collections still play a central role. While being informed by the models of participation in urban planning (Arnstein, 1969), sociological surveys of cultural participation (Morrone, 2006), or participation design in actual museum practice (Simon, 2010), our theoretical approach employs the access-interaction-participation (AIP) model as elaborated by media and communication scholar Nico Carpentier (2011). Distinguishing between access, interaction and participation, prevents these concepts from merging into one general concept of 'participation'. Within the framework of that model, access enables audiences to receive, whereas interaction is a more two-way process covering the exchange of meanings as well as their collective/social consumption. Most importantly, Carpentier integrates democracy theory with the AIP model so that participation can consequently be conceptualised as either minimalist or maximalist. Minimalist participation relies on the assumption that the political does not necessarily reach beyond the realm of conventional politics, and that professionals should be in control of the structure and processes, allowing them to homogenise audiences whenever necessary. The maximalist approach to participation, however, is based on a belief that the political is an underlying dimension of the social and that participation (ideally) entails power sharing, heterogeneity of audiences and also allows for structural changes (Carpentier, 2011: 17–22, 69).

How can minimalist and maximalist public participation processes manifest themselves with regard to museum collections? Taken in the context of museums and collections, minimalist participation affects content by contributing new objects and descriptions. Maximalist participation can also influence the structure of legitimation of museum collections, for example by negotiating and potentially diversifying the practices of accessioning and using museum collections. In order to bring the model closer to the traditional museum context involving material objects, participatory interventions can be theorised as having either 'virtual' or 'physical' influence on collections. The 'virtual' here refers not to online media but rather to an interaction that concerns, but does not physically affect, the museum collection, for example it leads neither to accessioning nor changes in the catalogue (including the digital catalogue). The 'physical' consequently refers to the participatory processes by which collections are physically affected, leading to a change in the collections: manifesting, for example, in new contributions and interpretations being accessioned. In both the virtual and the physical, the democratic dimension of the participatory process can be either minimalist or maximalist. In the minimalist participatory mode, museum professionals share some of their power with the active public when appropriating objects/meanings into museum collections.

Decision-making regarding exhibitions, cataloguing and collection usage remains in the hands of museum professionals. In the maximalist participatory mode, these processes can be negotiated with participants and collections might become structurally reorganised to some extent. Not only are new information or objects acquired, but particular museum collections and/or metadata can be reworked or redesigned, with notable participation by communities outside that of museum professionals (the theoretical framework for this is summarised in the Table 1.)

Table 1: Audience participation modes as manifested according to their influence on museum collections

	'Virtual' access/interaction	'Physical' access/interaction
Minimalist participation	Audiences contribute objects for display or discussion on-site/online, but these are not accessioned by the museum collections. Museum professionals decide which objects or meanings 'they are looking for' and will display	Accession of new objects or meanings created as part of a participatory process. Museum professionals decide which objects or meanings are to be accessioned and how

3. The Estonian National Museum and the formation of its collections

The ENM is a typical European ethnographic museum; founded in 1909 at the peak of the national awakening it still has symbolic meaning. As in many other European countries, the primary role of the museum was seen in preserving the old, fading peasant culture. Immediately after the museum was founded, the systematic collection of objects and oral history began, engaging volunteers across the country (Õunapuu, 2009: 46). During the Soviet era, the collections were divided into different museums, leaving 'ethnography' officially the main focus of the ENM. Museum collecting professionalised, its focus expanding to modern times and the public kept donating objects to a certain extent throughout the Soviet period. In the early 1990s, when Estonian independence was restored, a permanent exhibition entitled "Estonia: Land, People, Culture", focusing mainly on 19th century peasant culture, was opened. Popular understanding of the museum's collections and of the nature of museums is most influenced by the exhibitions on display, and above all by permanent exhibition.

Despite the public image, the current collection strategy deals primarily with collecting the everyday material of the Estonian people from the 1970s onwards, as well as objects from other Estonian-related peoples and communities, even though it is still common for the public to offer items related to pre-industrial peasant culture (Reinvelt, 2008). Historically, the ENM has looked at its public as more than just audiences for their exhibitions and customers for their services. Because of its research approach rooted in ethnology and folklore studies, individuals and groups have been considered subjects of study and 'informants', arguably becoming co-creators of the collections. In order to encourage the public to participate in the formation of collections, the museum clearly needs to provide some guidance on what content would be appropriate. Decisions leading to a growth in object collections are always made by museum professionals (confirmed by the chief treasurer) most commonly through fieldwork, although sometimes by purchases or public calls for donation. In recent years collections from smaller museums have been integrated to the ENM's collections.

In addition to object, photo and film collections, the oral history archive is one of the museum's largest. The majority of archive materials are collected oral history descriptions from fieldwork, and written contributions and donations from the correspondent's network, which was formed in the 1930s and currently has 300 members. As of the 2000s, the archive collections have also received materials from various participatory actions, the influence of which is discussed below.

4. Participatory initiatives and interventions at the Estonian National Museum

4.1. The Correspondents' Network: setting the "rules of participation" at the ENM[1]

Correspondent's Network is one example of a minimalist participatory initiative in which the museum invites, and also guides and controls, audience contributions. Every year the museum prepares a call for stories on subjects deemed important to the current museum research topics or exhibition production. The call is open to anyone, although the regular participants are mainly elderly people from different parts of Estonia who have time to contribute to subjects important to the museum. Through this network, thousands of pages of written responses (sometimes accompanied by photographs, films and objects) are accessioned by the museum every year.

4.2. "Donate a day to the Museum": the effective expansion of contributive participation

In order to build on the tradition of the correspondents' network, and to encourage other groups to contribute to the archive, a participatory intervention was designed for the ENM's 100th birthday, the 14th of April 2009. The date was used as an opportunity to make a mass appeal for 'donations' on that day. Descriptions of a 'typical day' in writing, through pictures or video resulted in 450 diverse descriptions. One of the project's breakthroughs was that the contributions came mainly from younger people.

People were sharing and producing new personal material with the museum willingly to give new knowledge to the museum. As with the traditional correspondent's network, this physical minimalist participation engaged new audi-

1 http://www.erm.ee/?node=57.

ences, but did not affect the formation of museum collections or professional practices. The call itself left the content, medium and discourse of the contribution relatively open.

4.3. "Estonian Moments": a civic initiative for crowdsourcing photographs online

In 2006 a young man named Tõnis Kärema contacted the museum and suggested that he voluntarily set up and run a webpage in order to help the museum collect photographs representing contemporary everyday life. The web-based database for collecting photographs that resulted, called *Eesti Hetked* (Estonian Moments),[2] has a number of similarities with the first 2 minimalist participatory processes described above – with the important difference that it is rooted in a civic initiative rather than one from the museum. The website is similar to the photograph senders' network, which includes annual themes on which the museum would like to see contributions (including an open theme). Consequently, the initiative was assimilated into that accession system.

At the beginning of the process, museum professionals expected that participants would contribute materials that were less relevant to the museum's preferences and that many contributions would be unsuitable. In actuality all those who complete the museum's traditional metadata slots (theme, time, place, action) have their photograph accessioned, except in the very few cases where the material violates personal rights.

Essentially, the museum collections appropriated a civic initiative and applied the same accession standards that apply to its own professionals. On the one hand museum professionals consider the public to be experts in their own lives, and therefore qualified to contribute, while on the other hand the museum is critical and hesitant when it comes to physical objects and the idea that everyone's physical interpretations should have a part in creating collections, as the next action shows.

4.4. Creative content from traditional audiences through an online competition

The purpose of the intervention entitled "My Favourite from the Estonian National Museum Collections"[3] (2010–2011), was to study communication between the museum and handicraft makers, who are probably the greatest groups of museum object collection users. While giving them the possibility to interpret the collections, it was also possible to study their perception of these collections. The competition took place in cooperation with handicraft makers' web communities,

---

2 http://www.eestihetked.ee/.

3 http://www.erm.ee/?lang=EST&node=1943&parent=252.

to whom online competitions (making handbags, gloves, etc.) have been regular practice. In addition, the possibility of offline participation was offered in order to include all handicraft makers. The competition was to choose an object – physical or digitised – and either, (1) make a copy, or (2) use the original for inspiration and create a new version of that object. The eventual choices reflected the most used collections: ethnographic textiles through the reuse of their ornaments.

Through the competition, 54 objects from museum collections received a personalised context from professional or hobby handicraft makers. The competition jury was elected by the web handicraft communities and by the museum, which both elected 5 experts known and appreciated for their quality of work. This and the museum experts' role challenged the participatory process, as one participant said: "Well, perhaps they would submit such work to an *Isetegija* [the handicraft makers web community] competition, however they don't dare to send it there [the ENM competition]" (Teppor, 2011). So some people were discouraged as they felt they were not good enough to participate. The jury selected 20 works for exhibition based on innovation and quality of technique, with the 4 best objects donated by the authors to, and accessioned by, the ENM.

Intervention in the museum's collections and giving new meaning to 54 objects falls within the sphere of reinvention techniques, such as stories relating to object manufacture, the meanings of objects and their new use; the results were seen by the collection department as material that was part of the participatory project, although it was not for inclusion in the collections. Subsequent discussion resulted in a picture of the copy and text relating to its manufacture being added to the museum's digital database, in turn being related to the original object, with the collections becoming physically affected as a consequence of this minimalist participation.

4.5. Hesitant steps towards an audience commentary board for the photographic collection

In 1993 the ENM's photo conservator Jüri Karm compiled an exhibition to introduce the museum's vast photographic collection, at the time numbering almost 200 000 photographic objects. In order to obtain a statistically representative sample a reproduction of every 182nd photograph (by order of accession) was taken from the collection, resulting in a total of 1080 reproductions being exhibited under the title, "With 1000 Steps..." The exhibition was displayed again in 2009, this time with an invitation to visitors to add free-form comments, an opinion, or a meaning to a photograph of their choice as well as the possibility to interact with the comments of other participant. A simple technology – post-it notes and pens – was used to enable people to post their comments on the photograph frame or on the wall next to the picture.

Through the intervention, an onsite feedback channel was opened for the visitors to share their opinion and virtually edit museum content (in the form of the photograph descriptions).

The over 80 posted comments given over 1.5 months was not an overwhelming number, but nevertheless indicated that there are visitors who are ready to participate, not only by 'rating' pictures but occasionally also by attempting to initiate debate, interacting with other visitors' comments or correcting captions.

A couple of shortcomings arose from a participatory perspective. As a result of the experimental nature of the intervention, what was missing from the framework was a clear message regarding how these comments would be used by the museum. There was virtual access and some virtual onsite interaction with collections in the exhibition space, but at the same time a considerable ambiguity regarding how participants' decisions could physically influence museum photograph collections and photograph catalogue descriptions. An implicit challenge here is how to foster trust among museum professionals towards visitors-as-participants in such activities.

4.6. Opening curatorship, diversifying collections?

The "Create Your Own Exhibition" project was launched in winter 2010 through a public call, inviting people to submit their exhibition proposals. While the agenda was to include new audiences in the museum's content provision, there was also an exclusionary mechanism to attract more non-professionals: museum professionals were not allowed to participate. Extensive public communication was followed by a new decision-making model in the form of an online and offline public vote open to all. Each entrant could compete in two categories an exhibition using the participant's own materials, and an exhibition using museum objects.

By June 2011, both of the winning exhibitions had been displayed in the ENM exhibition hall. The intervention's aim was to facilitate participation by attracting visitors to contribute a whole exhibition idea and actively engage in producing it. With rather low-level participation on the museum side, what took place was a maximalist audience participation.

The "Create Your Own Exhibition" project was first and foremost aimed at the exhibition hall, but which – like commenting on the photographic collection – also had implications for the collection. The intervention process was an invitation to choose and interpret (in one of the two competition categories) something from the museum collections. When the overall winner expressed the wish to donate the objects from her collection of funeral culture to the ENM after the exhibition, the challenge of whether the objects were legitimate for accession became eminent. Respecting on the one hand the integrity of the amateur collection and the principle of not overfilling museum collections with similar objects on the other, the 'gatekeepers' of the ENM collection and archive only accessioned a few objects, with the majority of objects (funeral song sheets, photographs) eventually acquired by the Estonian Literary Museum. Apart from these few objects, only photographs of the exhibition layout went into the ENM photographic collection, forming part of the routine documentation of the exhibitions within the museum. The maximalist audience participation did physically affect museum collections, but remained very small in scale and created no substantial structural diversity.

4. Audience participation and museum collections: getting physical?

A history of at least 80 years of inclusive participatory activities shows that making heritage production, in the form of museum collections, more inclusive and democratic is not particularly new or utopian at the Estonian National Museum. Ethnographic museum collections have traditionally been formed through contributions from the people, although in recent years new participatory projects have been implemented and existing ones refurbished.

Table 2 summarises what could be characterised as the beginning of the expansion of traditional minimalist participation, given with the number of objects involved in the exhibition space (virtual) and/or accessioned to museum collections (physical).

Table 2: Audience participation interventions at the ENM according to their influence on museum collections

	Virtual access/interaction	**Physical access/interaction**
Minimalist participation	"With 1000 Steps..." (80 comments) "My Favourite" (50 objects)	Correspondents' network (6193 pages in 2011) Estonian Moments (1000 photographs since 2006) "Donate a Day" (450 descriptions) "My Favourite" (4 objects) "Create Your Own Exhibition" (6 objects)
Maximalist participation	"Create Your Own Exhibition" (2 exhibitions, 33 proposals)	

When looking at the increase of participatory initiatives on the one hand and their influence on the museum collections on the other, what becomes evident is that the audiences are most active in minimalist physical participation, the practices of which also have the longest history at the museum. At the same time those participatory initiatives that don't fit in with legitimate accessioning procedures (both technically and in terms of decision-making) reveal that the structures underlying these procedures are still hegemonically dominated by the traditional expert-led system of taxonomy and thematic control over content. Pointing out such issues attempts neither to undermine the encouraging fact that increasing interaction with collections is possible, and not only with traditional audiences but also new ones; nor does it suggest that dedicated expertise should be totally deconstructed. The museum should investigate, though, the implications of participatory initiatives to the museum collections. What could be achieved by consciously using interventions to experiment with the 'physical maximalist' realm that would have structural influences on (part of) the collections? Perhaps it requires too much effort to 'defend' the collections against the donation of 'unwanted' objects. Museum professionals would also be required to recognise the potentially democratic nature of participatory processes and consequently take part actively in inventing alternative accessioning procedures/standards as well as ways to bring participation into the physical dimension structurally.

As of today, adding meanings 'virtually' in the exhibition space is not a completely alien process to the ENM, but without the continuous personal effort by someone from within the museum these meanings are in danger of remaining ephemeral and temporary. At the moment, only fragments of them achieve permanence by becoming part of the collections and thus there are practical implications in the 'virtual' realm of participation in museums. According to participatory logics, documenting and organising the new meanings created in participatory processes should not only increasingly be an integral part of the ENM's collections, but open new participatory horizons in the ways collections and audiences interact. One of the potential further developments of the intervention regarding the comments on photograph caption would be to create a dynamic, participation-driven description channel linked to museum object databases with the goal of pooling folksonomic metadata. Objects and meanings acquired through participatory processes could enable individuals and communities to build a more personal relationship with collections and be based on a platform for editing museum content both 'virtually' and 'physically'. Museum collections could thus become more dialogic and mobile than they currently are. However, when these new meanings and objects have instigated dialogue with museum collections and gained social relevance it should not automatically lead to restricted physical access because of the general preservation policy of the museum. In some cases, online opportunities could also enable users and communities to store the(ir) objects 'at home', but have the collection digitally integrated with the museum collection information system, thus relieving the pressure on museum's physical storage capacity. Such an inclusive and participatory "third collection", echoing the ideas of Homi Bhabha (Rutherford, 1990), would ideally be more open and user-driven. Even then, museums themselves will remain perhaps the most important stakeholders among these users by possessing the expertise necessary to help professionally preserve and contextualise the objects chosen to be part of collections.

Acknowledgements

The publishing of this article was supported by the Estonian Science Foundation grant no 8006 and the Estonian Ministry of Education and Science target financing projects no SF01800002s07 and no SF0180017s07.

Sources

Teppor, Marke (2011). Research interview with a participant of "My Favorite from ENM Collections" no 1.2, woman, 45.

References

Appleton, Josie (2007). Museums for 'the People'? In Sheila Watson (ed.). *Museums and Their Communities.* London and New York: Routledge, pp. 114–126.

Arnstein, Sherry R. (1969). A Ladder of Citizen Participation. *JAIP*. Vol. 35, No. 4: 216–224. Available at: http://lithgowschmidt.dk/sherry-arnstein/ladder-of-citizen-participation.html (accessed August 2, 2011).

Baddeley, Claire (2009). Managing the 'New Museology': The Changing Role and Management of Australian State Museums Since the 1980s. *The International Journal of the Inclusive Museum*. Vol. 2, No. 2: 89–97.

Carpentier, Nico (2011). *Media and Participation.* Bristol, UK & Chicago, USA: Intellect.

Knell, Simon 2007. Museums, Reality and the Material World. In Simon Knell (ed.). *Museums in the Material World.* London and New York: Routledge, pp. 1–28.

Morrone, Adolfo (2006). *Guidelines for Measuring Cultural Participation.* Montreal: UNESCO Institute for Statistics.

Reinvelt, Riina (2008). The Collections of the National Museum and Stereotypes. In Eva Fägerborg, Elin von Unge (eds.). *Connecting Collecting*. Stockholm: Samdok, Nordiska Museet.

Rutherford, Jonathan (1990). The Third Space. Interview with Homi Bhabha. – Jonathan Rutherford (ed.). *Identity: Community, Culture, Difference*. London: Lawrence and Wishart, pp. 207–221.

Simon, Nina (2010). *The Participatory Museum.* Santa Cruz: Museum 2.0.

Srinivasan, Ramesh; Robin Boast, Jonathan Furner, Katherine M. Becvar (2009). Digital Museums and Diverse Cultural Knowledges: Moving Past the Traditional Catalog. *The Information Society.* Vol. 25, No. 4: 265–278.

Õunapuu, Piret (2009). Mõte sai teoks. [The thought become real.] In Piret Õunapuu (ed.). *Eesti Rahva Muuseumi 100 aastat.* [100 years of Estonian National Museum.] Tartu: Eesti Rahva Muuseum, pp. 38–101.

DIGITAL TECHNOLOGIES FOR PARTICIPATION

Digital Cultural Heritage – Challenging Museums, Archives and Users

Pille Pruulmann-Vengerfeldt

Agnes Aljas

1. Introduction

Several studies claim that today's museums think too little about who the users of their online sites are, why the users go to these sites and how museums could better adapt the sites to their needs (Farber, Radensky, 2008; Roberto, 2008; Salgado, 2008; Samis, 2008). At the same time, there is a strong consensus that online space is very important in providing the pre- and post museum visit experience (Filippini-Fantoni, Bowen, 2007; Fisher, Twiss-Garrity, 2007; Durbin, 2008). In addition to extending the museum experience online, Estonian museums are facing the task of digitising increasing numbers of artifacts (texts, photos, films, objects, etc.) in order to place them in online digital storage spaces. According to the Estonian Digital Cultural Heritage Strategy (Eesti…, 2003) the aim of the memory institutions is to transfer cultural heritage in an uniform way to (almost) everyone, widen and expand the user groups and introduce Estonian cultural heritage outside the state borders and language space. As we learn from the Estonian perspective, cultural heritage is defined by state apparatuses and official institutions, by administrators and cultural engineers, whose task is to reproduce national culture and promote the identification of citizens with that culture. In most of the cases this is done in line with Bendix (2000: 38) who says that heritage can be distinguished from other ways of aligning the past with the present by, "its capacity to hide the complexities of history and politics".

Museums are facing many challenges connected with digitising their materials. In many ways, these challenges correspond to those that museums have faced for centuries. The focus is on the interrelationship between the users and the museums' collections; modern technologies are only one possible intermediary for these relationships. The classical roles of the museum are collecting, preserving, research and basic interpretation. In general, museums, especially if they are publicly funded, are seen as being obliged to give things back to society in order to "justify their existence", and according to Fleming (2007), this could be seen as the social responsibility of the museum. For Fleming, this responsibility is met when staff commit themselves to identifying and meeting the needs of the public, and when they place this at the head of their priorities (Fleming, 2007). Digitisation and making cultural heritage materials available online as subscribed to by the Estonian Digital Cultural Heritage Strategy (Eesti..., 2003) could be seen as one possible way of taking care of those responsibilities. In a country where public services are increasingly provided in online environments, museums face similar pressures from users and administrators.

2. Research design

This article aims to gather several data sets in order to understand the dynamics that exist between employees of the cultural institutions as providers of digital content, and youth groups as potential target audiences. The article starts by questioning the notion that there is an average internet user through survey data. By identifying and drawing on six internet user types, we continue to compare the insights from statistical analysis to materials from qualitative interviews. In 2008 12 semi-structured interviews were conducted in 4 different memory institutions in Tartu[1] (the second biggest city in Estonia) with the aim of opening a discussion about digitising and communicating cultural heritage. In addition, 2 focus group interviews[2] were conducted with young people (one with secondary school pupils and another with university students and young researchers) in order to find out how they would like to use cultural heritage.

For the memory institutions the digitisation of cultural heritage materials is seen to fulfil three basic needs relating to memory institutions: it serves as an aid to preservation; as way of opening access to wider publics; and as a way of inviting audiences to become active participants in introducing, learning and being involved with cultural heritage, either through the given interpretations or by inviting the community to give their own meanings to the cultural heritage materials stored in the museums. At the same time for young audiences, cultural heritage in general is seen as necessary for understanding both the past and also collective memory, mainly in the context of research projects and school papers. The role of memory institutions is seen as the systematic safe keeping of heritage for future generations, and thus is in line with the first two aims of the memory institutions' digitisation projects, although much less in accordance with the participatory focus.

Figure 1 gives an overview of the research design. We start by analysing in-

1 Four interviews with employees of the Estonian National Museum – refered to as ENM, six with employees of the Estonian Literary Museum – refered to as ELM, one interview with an employee of the Estonian Sports Museum – refered to as ESM and one interview with an employee of the National Archives of Estonia (1) – refered to as NAE.

2 Refered to as 1 for the pupil's focus group and 2 for the students and resercher's focus group, F or M for the speaker's gender.

ternet use, as in many ways how the internet is used reflects how the internet is conceptualised. We then use statistical analysis of the general population as a backdrop for more focused study of heritage professional's internet use. In this way museum workers' basic internet use is reflected in how they provide cultural heritage for the general population, and vice versa. As young people are generally seen as the key target group for online heritage, mainly because they are future users, online tools are seen as a way to foster interest in heritage among them. Thus we look at how the practices of internet use differ between the key groups.

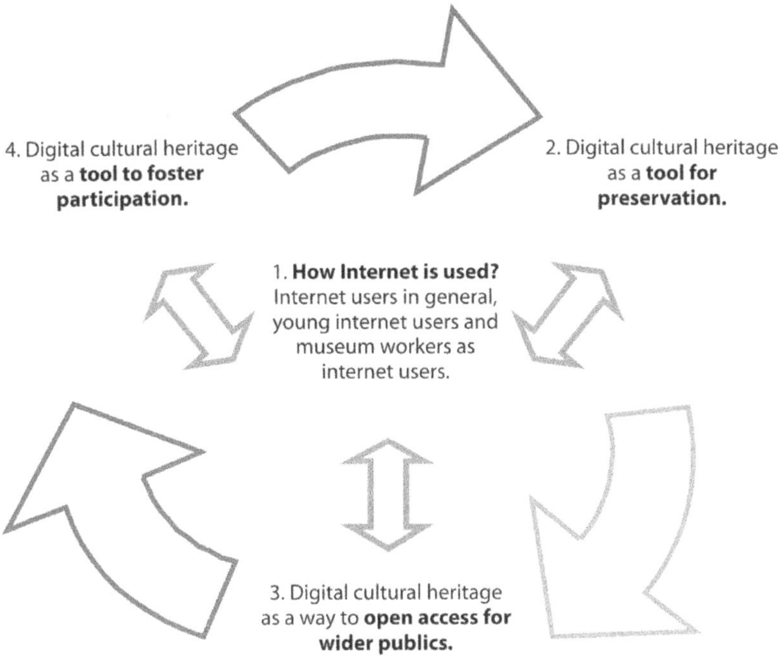

Figure 1: Overview of the research design

We will also look at the three key aims that memory institution have set as targets in the digitising of their collections. Online access and digital collections are supposed to help with preservation, open access and fostering participation and so the article also analyses how well these aims are met. The article concludes with some general considerations drawn from the discussion section.

3. The average internet user and the museum

As many authors have analysed, it is very important to understand the internet user within the museum context, as the online representations of a museum gives a very important pre- and post-museum visit experience (Filippini-Fantoni, Bowen, 2007; Fisher, Twiss-Garrity, 2007; Durbin, 2008). In order to give an empirical description of the 'average' internet user, we use data from the University of Tartu survey *Mina. Maailm. Meedia* (Me. The World. The Media) (2008). The survey was conducted in cooperation with a survey company and composed by a research team from University of Tartu. The representative sample consisted of 1507 people aged between 15 and 74 and it enables us to describe inhabitants of Estonia based on their internet use practices and to have some insight into their attitudes towards digital culture and content creation. Through cluster analysis, we have reached six basic internet user types who are similar to those described in our previous studies (Runnel, Pruulmann-Vengerfeldt, 2004; Pruulmann-Vengerfeldt, 2006; Runnel, Pruulmann-Vengerfeldt, Reinsalu, 2009). The types have remained fairly stable (Kalmus, Keller, Pruulmann-Vengerfeldt, 2009), thus enabling us to make assumptions on future internet use as well.

In general, the types can be divided into two broader types, each of which has three types. On the one hand more active internet users, (varied use, practical and pragmatic information-based use, entertainment and communication-oriented use) and, on the other hand, three types of more passive internet users. These types include users oriented towards information and entertainment as well as infrequent users who come into contact with the internet so rarely that it is impossible to distinguish clearly developed practices of use. Figure 2 provides an overview of the online activities of the user types, comparing their frequency of engagement in the most popular activities and activities related to digital culture. Figure 2 gives a comparison of internet users according to the nine most distinguishing internet user practices. Respondents rated on a scale of 1–7 how important this activity was for them. On average, the responses ranged from 2–4, depending on the activity. In the figure, one can see the variation from the average, marking how much this particular activity was considered more/less important than the average response.

Active, versatile internet users (14% of all internet users) are more active with regard to all manner of internet use compared to the other groups. For them, the internet is an environment where they satisfy their need for information, entertainment, belonging and participation (Figure 2). This type includes a greater proportion of women, people aged 20–39, and people with a higher education. Together with the next internet user type, they are actively contributing online content. Uploading photos is the most common activity where content is provided in the online environment and social networking sites come second in online content creation practices.

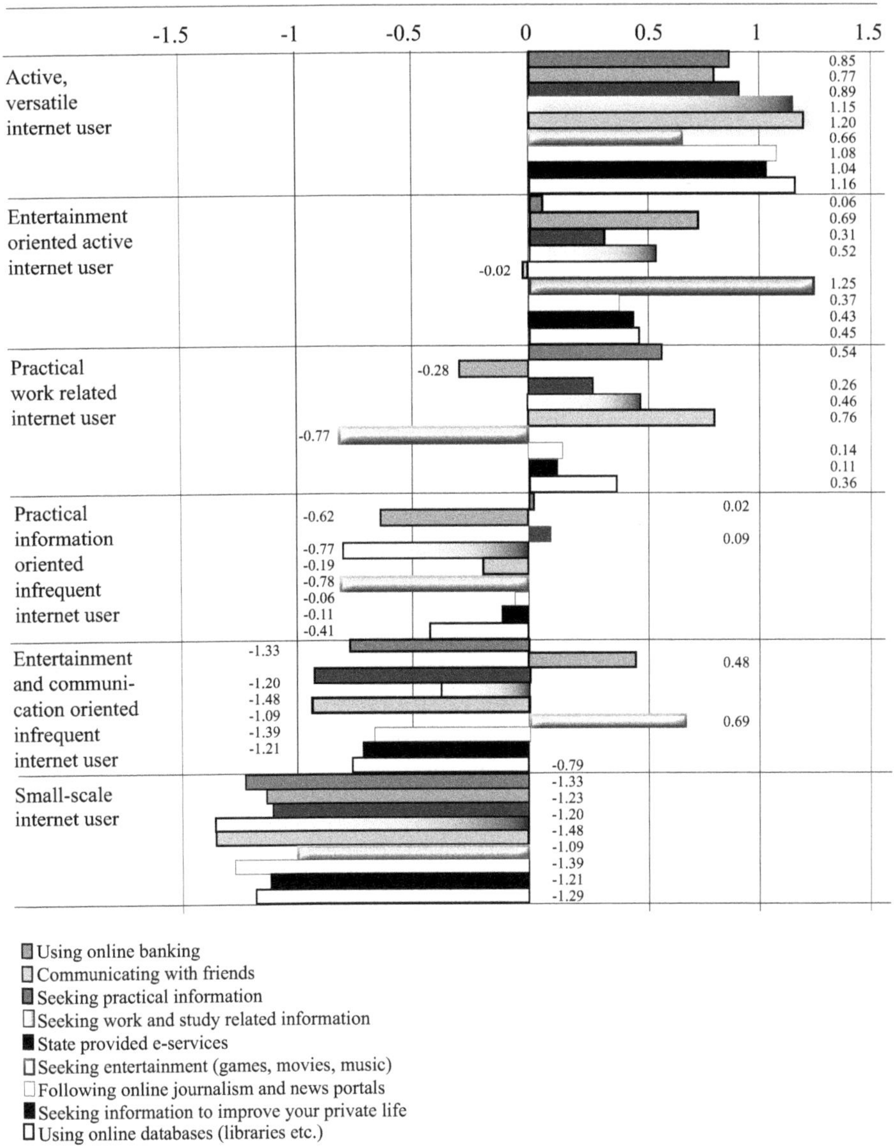

Figure 2: *Nine online activities distinguished by internet users types (numbers indicate deviation from the mean scores of the average users).* Source: Mina. Maailm. Meedia. (*Me. The World. The Media.*) 2008

Entertainment oriented active internet users (20%) concentrate mainly on searching for entertainment, watching/listening to TV and radio shows through the internet, and also on the consumption of culture. This type of user is generally active, however, and tends to search for information and use the internet to gain access to practical services if necessary. This user type includes people who consider it important to participate in blogs and forums. The largest number of internet users of this type belongs among the 15–29 age group, with the Russian-speaking population being represented slightly more among the entertainment-oriented active users. This group is most active when it comes to different forms of content creation. They upload photos and videos, and participate in forums and social networking sites.

Practical work related internet users (22%) focus primarily on information and practical activities, in addition to being significantly more active than average in using e-services. Their online communication is mainly work related and considerably less personal than that of the average internet user. They also search for significantly less entertainment than the average internet user. This group is dominated by women, people aged 30–49, people with a higher education and members of the Estonian-speaking population. In addition, people belonging to this group are more likely to have an average or high income.

The largest group among the passive internet users is the *practical information oriented infrequent internet users* (20%). Their internet use is characterised by a somewhat higher than average use of information and slightly higher than average use of online banking solutions. At the same time, the variety of their online activities is somewhat larger than that of the next user type. This group more commonly includes older people, women and people with a secondary education.

The online activities of *entertainment and communication oriented infrequent internet users* (15%) are characterised by searching for entertainment as well as communication with friends and acquaintances, while their internet use remains passive with regard to other purposes. This group includes more men, members of younger age groups and therefore also people with a basic education and those belonging to the lowest income group.

Small-scale internet users (10%) are not characterised by any specific practice of internet use and their online behaviour is generally in the developmental stage. Infrequent users comprise a larger than average share of older people and people with a secondary education, as well as members of the Russian-speaking population. They also are the least active when it comes to contributing online content.

Figure 3 relates these internet user types to their attitudes towards going to a museum. In general, it illustrates the idea that the more active people are in their attitudes towards life, the more frequent internet users they are as well. Although one might assume that if a person holds traditional values and a conservative attitude, they would rather go to a museum and not use the internet that much, the research illustrates that this assumption doesn't hold true. The more active internet users are, the higher is the chance that they will also be more likely to go to the museums.

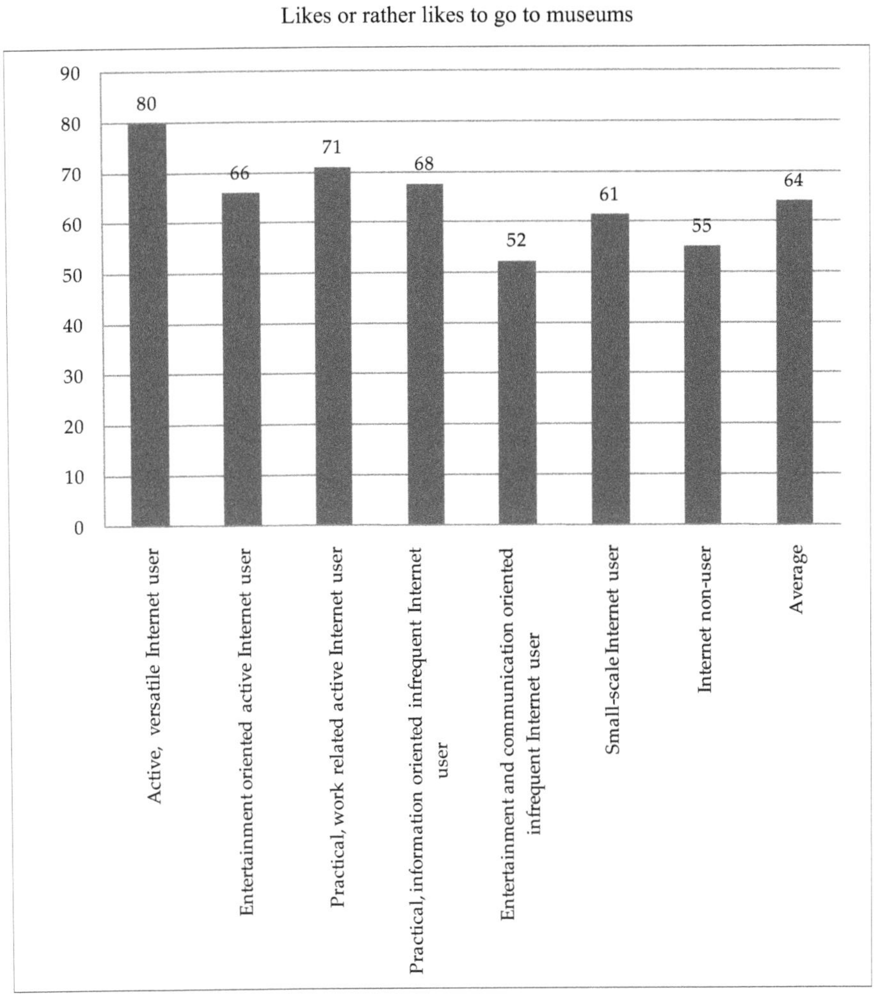

Figure 3: *Percentage of Internet users who like, or like very much, visiting museums.*
Source: Mina. Maailm. Meedia *(Me. The World. The Media) 2008*

4. Internet experience and the practices of memory institution employees in contrast with the practices of young people

When compared with young people, museum and archive workers tend to belong to the practical work oriented groups. Some of them can be classified as active internet users, while others are more infrequent. Overall, their use is very much oriented towards getting things needed for work done and much less towards entertainment related or leisure use.

For employees of memory institutions, their everyday work and most of their day is spent at a computer and on the internet. A museum employee's day often starts with reviewing and answering e-mails, and a large part of their professional communication takes place via e-mail and internal websites, which have made the sharing of information easier. E-mails have made communication and exchange of information more active and operative in Estonia and abroad. At the same time it is stressed that direct communication is still important in the functioning of an organisation and plays a significant role in developing further web-based communication outside the institution.

> By and large, I get all the information I need for work [from the internet], although we do have department meetings, but I also get the information I should know [from the internet]. /.../ On the internal web I can express my opinion and communicate with colleagues, that probably joins it all up and enables me to promptly use information and everyone to look at one and the same thing, increases and enhances the quality of work. But I still think that we also need these joint meetings. (ENM[3])

First of all, the internet is used for finding work-related information. The homepages of the Ministry of Culture, Tartu City and museums are used most often to find necessary information, contacts and documents. Database search systems are important in everyday work; depending on the nature of the work, the databases of the Institute of the Estonian Language, the Estonian Literary Museum and the National Archives of Estonia are used, as well as the library database ESTER, the Amazon bookstore and various dictionaries. People mainly stay in the Estonian-based internet space, venturing into foreign language web-space seldom and then rather out of interest than everyday need. Finding and reading important speciality articles in internet databases has become important for people who are proficient in the English language.

The internet allows people to be up to date with the activities of memory institutions around the world. Employees often visit the homepages of professional unions (e.g. ICOM) or museums and archives in Finland, Sweden, Denmark and

3 The interviewed experts are quoted with reference to their institution.

America, being most interested in novel solutions and gaining inspiration for professional activities. The interviewees hold in high regard databases of professional importance, which enable the necessary information to be found without entering the research hall or library. As a significant factor in using the databases, the interviewees mentioned user friendliness, which for them means the simplicity of navigation in the search system and the speed of finding the required information. If the search system of a database is too complicated or the sought information is not found, the database is not used again.

Outside working hours, the interviewees primarily use the internet to find information. The internet provides help finding cultural events and weather forecasts and assists in making travel plans; people also use it to read the news, use banking services and the electronic school portal for checking children's progress in school. A common opinion is that making big purchases has become easier via the internet. Internet portals are the main source of assistance in buying a car or property and making price comparisons.

Besides the use of e-services, finding hobby related information is important for museum employees. People follow thematic blogs and homepages where they read the news and look for answers to specific questions. The internet is not considered an important place for personal communication or entertainment. Work related communication has moved to the internet and therefore people tend to prefer direct communication outside working hours.

> For very many people, everyday life has moved to the internet. For me, it has not moved to the internet – some parts have, and I cannot say when it should happen that my everyday life will move to the internet. (ENM)

As the employees of memory institutions generally fall into the categories of practical information oriented internet users, both more active and infrequent, we asked them who they considered to be 'ordinary internet users'. The answer was a vague description of 'ordinary' in which the users referred to younger people who use the internet more actively and largely for entertainment. The interviewees thought that young people use the internet for communication, sharing information and photos with friends, writing blogs and homepages, writing comments and watching films. The 'ordinary' users' skills in finding necessary information on the internet are better and they can also manage more complicated databases or search systems.

In our focus group interviews, the respondents could belong to the versatile active internet users, entertainment oriented internet users or in some cases, for more mature students, to the category of work oriented internet users. Most focus group members are in the active internet user category. On one hand, the internet is used for practical needs such as research or information searches, and on the other hand entertainment files such as films and music are downloaded, and the internet is used as a social networking environment.

> Orkut, YouTube and MSN – it's like a trio. And if I need to do a search, I use Google, not Yahoo, because I like Google more. And because I have the neti.ee portal as my browser start page, I often find myself on that site. (1M)

> Every day… Google and MSN, life is unthinkable without them. And of course, like anyone. /…/ Information search, entertainment, looking for applications to download from time to time… (2F)

> And I use the internet, too, like anyone else: to read e-mail and to look for all sorts of things and for watching all kinds of movies and… /…/. Oh, yeah, I keep a blog, too. For the reason that I want to see what it is, what it's all about. (2F)

Active users greeted technological innovations with great interest and tried them out to see if they were compatible with their user preferences. Databases or web portals from which it was complicated to find interesting material were usually discarded after an initial, disappointing, experience. Thus the process of finding information must be compatible with the user's existing browsing logic and user experience. Another aspect that is considered unsuitable with regard to use of a portal is if it is too time consuming to distinguish relevant information from irrelevant information, or if the pages are overloaded with banners and animated adverts.

> It may also be that some sites are really visually 'busy' and you practically can't understand where the things are listed. If there are many, many ads, or if... the information could be summarised much more concisely but it is all spread out. (1M)

In comparing the qualitative data to the quantitative, we can say that almost 40% of the general population of internet users fall into the category of work related users. When considering that, for them the internet is used for qualities that are mainly relevant for their work, we can see a gap forming. Browsing museum websites or searching databases for heritage information is work for very few people, thus making the key target groups for museum websites those who belong among the versatile active internet users and entertainment related users. Most of the museum and archive's websites and heritage databases are designed by people whose primary use is work related, while at the same time the primary target group uses internet for leisure and fun. This generates a situation in which there is a potential gap between the understanding and conceptualisation of the internet, and this might in turn lead to a usability gap. The three key uses that heritage institutions outline, and for which the websites and online databases are designed, will be investigated in the next sections of this article.

5. Preservation and publication of collections online

We presume that museums and their collections exist for their users and visitors. Museums have defined the needs of the public in the traditional context, and within known environments such as exhibitions, etc. However, in addition museums should think in the same way about visitors to their online environments. Many studies have indicated that museums do not try to understand database users when creating online databases about museum collections (Farber, Radensky, 2008; Roberto, 2008; Salgado, 2008; Samis, 2008). Very often the basic idea of the database is to create the web-based museum objects gallery. Similarly, Estonian museums and archives see the creation of improved preservation possibilities and the reduction of the damage caused by usage as the foremost objective of digitisation. The practice of digitisation has so far been focussed primarily on materials most used by researchers. Thus, one can see that in these cases, digitisation is very much a user driven activity. This kind of digitisation practice has enabled the National Archives of Estonia to claim that 90% of their most-used sources are available online. In addition, contract work materials are digitised on an ongoing basis for exhibitions and publications. Materials that the users have not yet discovered in a collection or not shown interest in are not a priority for digitisation, primarily due to the lack of the financial and time resources necessary for the process. Memory institutions also have fewer resources to focus on what Roberto (2008) and Samis (2008) have stressed as vital: that museum objects in the "web of data" should not only be information sources, but also offer interpretation.

> I think that much currently depends on financial possibilities. There are ideas and thoughts, and another thing is that people should cooperate in respect of financial possibilities. And cooperation between institutions inevitably takes time (ELM).

However, this kind of practice leads to an unsystematic and often project-based digitisation process. More often than not, the interviews indicated that digitisation is first and foremost seen as a technical process of generating digital files from documents, and much less attention is paid at the information architecture, interpretations and systematisation of these works. As a first step many memory institutions in Estonia have introduced a web-based ordering and delivery system, which requires a precise order from the client. This potentially makes user interaction with the collections easier, but also challenges them to have greater preknowledge of these collections.

So far, all the cultural heritage digitisation strategies have remained on paper and the lack of real cooperation between major institutions has also not enhanced uniform development. Various institutions have created several different databases from similar material, although these do not form an integral whole or make finding information from a single access point easier for users.

> Between archives, we have already learnt that users are not interested whether the thing they are looking for is in the state archives, history archives, film archives – users are interested in using the information (NAE).

Similarly, a shortcoming cited by focus group members, relating to orientation within memory institution databases, is the lack of a single unified system and the complexity of finding databases. In practice, finding and using many museum or archive databases requires guidance from a teacher or advisor because memory institutions lack visibility in search engine results.

> I was a senior year student in upper secondary school when ERNI was introduced and it wasn't really a finished product. My literature teacher demonstrated it. For me it was interesting but it was completely different and they were texts that I would not otherwise have read or viewed and it was very interesting. For me, it was a real eye-opener and in some sense I have been using these texts to this day. (2F)

This indicates that proper guidance to online databases can be inspiring for the user, but only a few focus group members have continued to use the databases they found. The use of the databases is made more difficult above all by a lack of knowledge about the content they offer, which makes it difficult to perform a search; moreover, the data structure is too complicated for consistent use.

In the web-based presentation of their collections, experts have so far given low importance to the desire to increase the openness and recognition of memory institutions, and therefore increase the number of users. Facilitating access to collections through web publication can be considered the second objective of digitisation. The superiority of the original artifact is still considered more important than the interpretations and value generated with the help of its digital representations.

> Users will definitely be glad if they can see it [data on the internet]. Because users are very lazy… we would, of course, like to see users checking out those things on the web and having access, but also coming here. I can understand users – archives are open on workdays and in working hours, and likewise all archives. One has to be retired, on childcare leave or unemployed to be able to go and study archive materials and original documents (ELM).

In many ways, museum and archive workers in Estonia still portray the object-centredness of the Victorian museum where viewing the glass caskets was more relevant than the experiences and relationships with the museum user and the artifacts. This is also reflected in the view that online databases are only incentives for the user to find their way to the original artifacts stored in the museum or archive.

6. Widening access through established, and potential new, databases

Depending on the topic of an exhibition the expectations of various target groups, and the relevant context, are taken into account and the ideal viewer of visitor envisaged quite clearly. The same can't be said of the online exhibitions or databases.

One objective of the databases so far created in the National Archives of Estonia has been to improve the availability of collections to hobbyists in addition to researchers.

> [T]he physical research hall in this building has approximately 20 workplaces and 40–50 people pass through there every day. Sometimes less and sometimes more. Now we have opened a virtual research hall and I think we will have about sixty users early in the morning [...] and at the best times we will have over 500 users simultaneously from all over the world. Archive using possibilities have increased tremendously (NAE).

Database search systems and the presentation of materials depend on the system of collection, while the meta-data added to this information is selected based on the needs of the database 'ordinary user'. In the context of databases, the term 'ordinary user' first of all means researchers of various levels and target groups with specific interests – teachers, students, the media and museum workers. The common assumption is that from the start these users are highly knowledgeable, motivated and interested in museums and studying cultural heritage via the web: if a museum loads something up, these users will come anyway.

Feedback from database users has so far been completely neglected and in the few instances some comments have been made, they are in general positive. This has given grounds to presume that the databases are user-friendly and that finding the necessary information is easy. There is no information about various user groups, and the following is a rather common answer:

> But we haven't received much feedback on who is the ordinary user of databases. It is clear that the media uses it, various portals, teachers – from them, we have received feedback – when they are asking whether they can use it or telling us that they found this or that fault and could we please fix it (ELM).

Cultural heritage institution professionals see that the web environment could bring people with no research or museum interest to museums, particularly the younger age groups. Digitised collections and search systems enable museums to attract interest and bring in wider user groups to view original materials and artefacts. Similarly, the participants in the two focus groups assumed that good and user-friendly databases would help bring them closer to the museums' activities.

Users were asked to describe ideal web portals that would draw them to museums, and five principles can be summarised from their discussions.

(1) A memory institution must have a presence on the Web along with all of its content, as often it is not possible for users to visit the institution.

Here the indication is that, although professionals would like to see online collections as leading to the physical museum or archive experience, the youth focus group participants see this as a less important factor. Museum professionals do not believe that users will completely lose interest in viewing originals because of digitisation. They are confident that no virtual exhibition or database can replace a three dimensional original copy or an old photograph, film or document. A digital database is seen as first of all an incentive to interest the user and spark the desire to see the original. At the same time the experts admit that many users will probably not make it any further than the databases. However, for the participants in the youth focus groups, museum databases should be able to sustain online representations on their own.

(2) A database must contain an introduction to its structure and data, and contain abundant illustrations, video material and interviews.

When digitising materials, museums often focus on one type of material at a time – for instance, all glass negatives (daguerreotypes) all maps, etc., while users would much rather have materials that are interlinked through a story. Here the digital museum can almost be described as undergoing a rebirth, in a fairly similar way to that in which the Victorian museum as a storage space of objects was reborn though Neurath's revolution in early 20th century (Henning, 2006).

(3) The data (i.e. list of sources, digitised sources) must also include interpretations, context and background information that would help create associations and create a whole, as well as containing links/references to other related databases.

While in the museum context professionals see digitisation as an aim of its own, and want to have the objects tell their own stories, young users are much more interested in having that work done for them through the provision of materials that are already interlinked and have interpretations provided.

(4) Multifaceted information should be structured pursuant to user profiles so that it is possible to distinguish between information that is relevant for researchers, and that which is relevant for users who simply wish to find interesting information, and so avoid information 'noise'.

Here the young focus group participants indicate a clear understanding of the differences between potential digital heritage material audiences. The possibility of differing user levels is somewhat distant from the heritage professional's view, in which all database users are perceived as professionals and equally interested and knowledgeable in all aspects. In our focus groups, two different potential audiences – secondary school pupils, and university students and young researchers – acknowledged that interests in different subject matters vary, and therefore the differentiation of user profiles seems like a good (albeit time- and resource-intensive) solution.

(5) Various cultural heritage databases should be consolidated in one environment and the structure should be unified.

The super-database of all Estonian cultural heritage materials seems to be a common wish for all – the bureaucrats who drafted the Estonian Digital Cultural Heritage Strategy, museum workers and potential users. However, today the lack of resources, strategic planning or a conceptualised understanding of public and museum needs stands in a way of this dream coming true.

7. Engaging users in the creation of collections

The third objective of digitisation is to engage users in the collection of digital materials and the creation of cultural heritage via the web. Henning (2006: 130) sees that the internet in its database-like structure would enable museums to re-enact the Foucauldian dream of the return of curiosity, and thus the age of curiosity cabinets from the history of museums. Yet in many ways despite the opening up, and participatory proclamations, of Estonian digitisation policies, in reality the digitisation of materials is ultimately focused on keeping the "Victorian era glass caskets", even though they are now in the digital form. Cultural institutions are still seeking solutions for participatory engagement that would satisfy all the parties. Although the most natural thing in Estonian digital space is online commentaries, and users are familiar with seeing them in variety of forms and environments, there is still a distinct disinterest in participating in the museum's activities. This is by no means helped by the fact that museums are looking for a quality of material that, for the professional, is not always reflected in those hastily scribbled remarks of the online commentary tradition. The high standards and strict rules applied to items normally worthy of museums' attention raises the entry requirements for participatory projects in some cases to unreachable levels.

> At one point we were having a whole lot of trouble with it; because spam robots discovered it and we had 300 comments along the lines of 'see beautiful girls here'. Then we solved it by restricting comments from abroad. /…/ But we did create the option, hoping that people will write down their customs. But we need to think about how to change it. Because back then it wasn't so common to comment on every article, saying that it is stupid. Today, this is much more common. (ELM)

Salgado (2008) and Farber and Radensky (2008) have shown in their studies that users are more prepared to interact with museums in the context of new technologies and web databases, but only if the systems are user friendly. So the most important factor in creating web databases are the understanding of the users needs and their potential motivations in using materials. It is also important to trust users and the public in creating new cultural heritage, and to help them establish an environment for communication. However, as Durbin (2008) has shown in her paper about Web 2.0, that modern online possibilities should not be viewed as not being technological or experimental, rather they should be seen as tasking museums to interact in new ways with the public.

Figure 4 gives an overview of how familiar Estonian internet users are with contributing content online. Two thirds of the most active online participants – versatile and active entertainment oriented users who we have referred to as the key target groups for the museum – have uploaded photos to the internet. This indicates that there is at least some willingness and habit to provide content in the online environment. At the same time, in the more passive groups, one can see that almost all content creation practices have been tried out by less than 10% of the group. In many ways, this can generate dilemmas for museums. Those who are more familiar with participating in the online environments may be seen as not so 'serious' in their internet use and thus also the content they contribute may be more entertainment related.

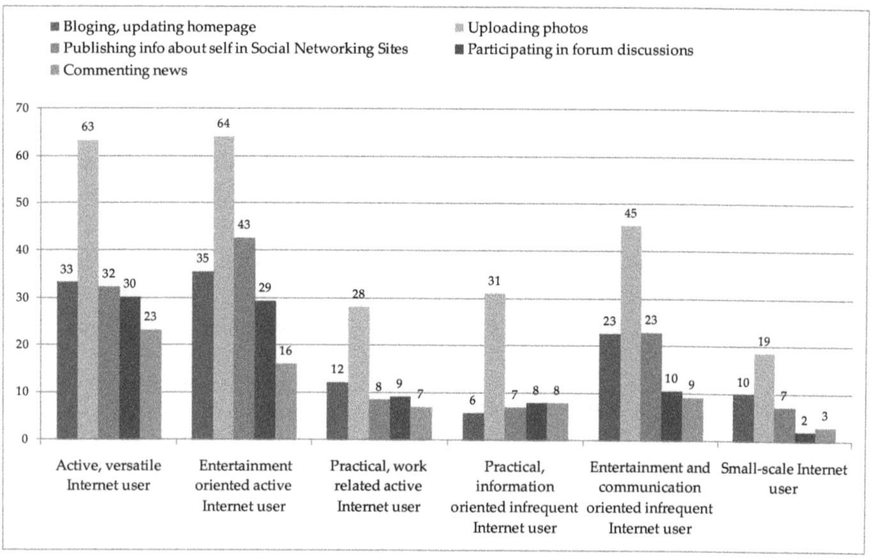

Figure 4. Percentage of internet users ever contributed to particular types of online content

All the professionals interviewed understand that it is of no use for a museum if users collect materials on their own, yet have no option to add them to the museum's collection. The creation of these possibilities requires changes to be made in the work organisation of institutions as well as separate management of materials and communication with users. At the same time, professionals expect that when users add materials to web-based databases, they must act in a way that is compatible with the institution's collection systems, i.e. be knowledgeable of cataloguing and meta-data information. Nevertheless, experts find that the collection of digital material has helped them to better understand users' needs and to observe and understand their activity patterns in the internet environment. In many cases, the interviewees thought that users have not yet developed the habit of contributing to memory institutions, and that at the moment electronic contributions have become less personal than information received in the conventional written form.

> Well, when we were collecting school heritage, it differed from 1992 most of all by the fact that [at that time] there was an option to reply electronically... People could get questionnaires both by e-mail and from the computer. But the material we received on paper was more properly and purposefully prepared, because anything can happen on the Net... People write a little bit and anonymously, but there is no anonymity on paper… if it is organised and assisted by teachers – you can't always check that with computers (ELM).

These contradicting results indicate that although there is a willingness and need to listen to the user as a source of modern heritage material, at the same time 'ordinary people' have a perceived distinct lack of the skill necessary to participate in museum activities. At the same time, some members of the general population have enough practice creating online content that, should there be incentives from the memory institutions to provide content, they might be able to do so. However, content creation practices are not overly popular and in cases where people might be interested in participating in a museum or archive's activities, they might not have the necessary skills. And if the 'quality' threshold set by the museums is very high this only increases the skills barrier even more.

8. Conclusions

The key gap between heritage websites and their uses potentially stems from the different user practices of heritage professionals and their target audiences. When people for whom heritage is mainly related to their professional activities – with all their long-standing professional practices – start designing online databases and websites for youngsters whose 'holy-trinity of the internet' are formed by MSN, Orkut and YouTube, then there is a strong potential for miscommunication.

In order to make web environments that are usable, the key is to understand the user's motivation for wanting access to digital heritage materials.

Every memory institutions sees its main role as storing and preserving its collections. Digitisation is one way of maintaining the ideal storage conditions for the museum or archive objects by making use of their digital copies, thus enabling the storage of the original. At the same time, all over the world, the user of the museum has been increasingly in focus and museums are becoming more and more user-centred instead of being centred on their collections. Creating and interpreting cultural heritage has been distanced from the experts and curators, and rather the community whose cultural heritage is at stake is seen as the main interpreter. However, the community does not always grasp this role. In our discussion with cultural heritage institutions' professionals and members of young audiences, who are foreseen as the key target groups for digital collections, it transpired that audience members are keen on searching through and looking at heritage materials, preferably across various collections, but they would rather have the interpretations with the material. While the technological opportunities, whether web 2.0 or another platform, are more and more readily available, the role of the user is as fuzzy for the Estonian museum and archive employees as it is for young members of potential audiences. It is often felt that we first have to sort out the data – digitise, organise, make available – and only then can we look at the interpretations.

The key focus of the interviews, both for professionals and users, was centred around making digitised materials available to users. This inevitably boils down to the question of maximally effective information architecture. With increasing amounts of information available online, both users and producers of online materials feel that the searchability, clarity and variety of information is vital. However, in many cases, museum and archive professionals feel that users should master the traditional practices of cataloguing and key-wording the artefacts rather than having the museums and archives adapt those to new conditions. Although no one assumes that cultural heritage must compete with social networking sites or YouTube, one should face the fact that memory institutions are seen as aspects of the entertainment sector and that young people today are first and foremost familiar with the aforementioned online environments. This poses a challenge for the memory institutions to grasp the possibilities offered by those online spaces, while still maintaining the traditional values and conceptions necessary for their professional identities. Many museum and archive experts feel that as existing cataloguing systems and database structures have worked for museums for nearly a hundred years, they should continue to do so. Others understand the challenge of opening museums up and the need to adapt to less experienced users' knowledge.

Despite the fact that traditionally the logic of different memory institutions differs – museums see their role as more focused on interpretations, while the primary focus of archives is one of storage and availability – users of heritage materials online do not care so much about the institutions' backgrounds. For them, the key concern is the availability of the materials and assistance that professionals can provide in interpreting these materials.

In conclusion, we can say that in many ways, the online spaces and databases of the museums and archives provide a multitude of challenges. The first role of digital cultural heritage is to aid the storage of artifacts and to save them for the future. At the same time, institutions are not that interested in updating their own cataloguing or meta-data processes and thus may miss out on the opportunity to increase the usability of the materials once they have been digitised. Thus digital collections may remain as unused and untouched as the originals in the vaults. Secondly, although the need for relevant and easy-to-use online spaces is understood, the underlying assumption is still that people need to come to the museum to see the originals, and not just make use of the digital copies. In seeing digital space as merely complementary to the 'real' environment, many good opportunities may easily be missed. Thirdly, there is a need for mutual education in order to increase museum and archive participatory possibilities, and therefore to grasp the potentials and opportunities hailed by new technologies. When museums see little value or relevance in user-provided materials, users will not easily learn to provide materials that are of interest for museums. Today, new technologies provide the potential to close the gap between memory institutions and the general population; however, unless there is a considerable change in the way memory institutions think about the audiences of the heritage, this potential may never be realised.

Acknowledgements

The publishing of this article was supported by the Estonian Science Foundation grant no 8006 and the Estonian Ministry of Education and Science target financing project no SF0180017s07.

References

Bendix, Regina (2000). Heredity, Hybridity and Heritage from one Fin-de-Siècle to the Next. In Pertti J. Anttonen et al. (eds.). *Folklore, Heritage Politics and Ethnic Diversity.* Botkyrka: Multicultural Center, pp. 37–54.

Durbin, Gail (2008). Just Do It! Web 2.0 As Task, Not Technology. In Jennifer Trant, David Bearman (eds.). *Museums and the Web 2008: Proceedings*. Toronto: Archives & Museum Informatics. Published 31.3.2008. Available at: http://www.archimuse.com/mw2008/papers/durbin/durbin.html (accessed February 13, 2009).

Eesti kultuuripärandi digitaalse säilitamise riiklik strateegia 2004–2007 [National strategy of digital preservation of Estonian cultural heritage.] (2003). Kultuuriministeerium. http://www.kul.ee/failid/DigiStrateegia_final.pdf (accessed February 13, 2009).

Farber, Allison; Paul Radensky (2008). Living Museum®: Supporting the Creation of Quality User-Generated Content. In Jennifer Trant, David Bearman (eds.). *Museums and the Web 2008: Proceedings*. Toronto: Archives & Museum Informatics. Published 31.3.2008. http://www.archimuse.com/mw2008/papers/farber/farber.html (accessed February 13, 2009).

Filippini-Fantoni, Silvia; Jonathan Bowen (2007). Bookmarking in Museums: Extending the Museum Experience Beyond the Visit? In Jennifer Trant, David Bearman (eds.). *Museums and the Web 2007: Proceedings*. Toronto: Archives & Museum Informatics. Published March 1, 2007. Available at: http://www.archimuse.com/mw2007/papers/filippini-fantoni/filippini-fantoni.html (accessed February 13, 2009).

Fisher, Matthew; Beth A. Twiss-Garrity (2007). Remixing Exhibits: Constructing Participatory Narratives With On-Line Tools to Augment Museum Experiences. In Jennifer Trant, David Bearman (eds.). *Museums and the Web 2007: Proceedings*. Toronto: Archives & Museum Informatics. Published 1.3.2007. Available at: http://www.archimuse.com/mw2007/papers/fisher/fisher.html (accessed February 13, 2009).

Fleming, David (2007). Are Museums about Stories or Objects? – *Interpretation Conference London, 24 September 2007*. Available at: http://www.bms.edu.lv/index.php?option=com_content&task=view&id=1 8&Itemid=32 (accessed July 28, 2009).

Henning, Michelle (2006). *Museums, Media and Cultural Theory*. Maidenhead: Open University Press.

Kalmus, Veronika; Margit Keller, Pille Pruulmann-Vengerfeldt (2009). Quality of Life in a Consumer and Information Society. In Marju Lauristin (ed.). *Estonian Human Development Report 2008*. Tallinn: Eesti Ekspressi Kirjastus, pp. 102–107.

Pruulmann-Vengerfeldt, Pille (2006). *Information Technology Users and Uses within the Different Layers of the Information Environment in Estonia. Disserationes de mediis et communicationibus Universitatis Tartuensis* 4. Tartu: Tartu University Press.

Roberto, Frankie (2008). Exploring Museum Collections On-line: The Quantitative Method. In Jennifer Trant, David Bearman (eds.). *Museums and the Web 2008: Proceedings*. Toronto: Archives & Museum Informatics. Published 31.3.2008. Available at: http://www.archimuse.com/mw2008/papers/ roberto/roberto.html (accessed February 13, 2009).

Runnel, Pille; Pille Pruulmann-Vengerfeldt (2004). Mobiilid, arvutid, internetid: Eesti infoühiskonna künnisel. [Mobile Phones, Computers, the Internet: Estonia on the Verge of Information Society.] In Veronika Kalmus, Marju Lauristin, Pille Pruulmann-Vengerfeldt (eds.). *Eesti elavik 21. sajandi algul: ülevaade uurimuse „Mina. Maailm. Meedia" tulemustest.* [Estonian Lifeworld in the Beginning of the 21st Century: Overview of Results from Research Project Me. The World. The Media.] Tartu: Tartu University Press, 147–162.

Runnel, Pille; Pille Pruulmann-Vengerfeldt, Kristina Reinsalu (2009). Estonian Tiger Leap from Post-communism to the Information Society: from Policy to Practices. *Journal of Baltic Studies*. Vol. 40, No. 1: 29–51.

Salgado, Mariana (2008). Breaking Apart Participation in Museums. In Jennifer Trant, David Bearman (eds.). *Museums and the Web 2008: Proceedings*, Toronto: Archives & Museum Informatics. Published 31.3.2008. Available at: http://www.archimuse.com/mw2008/papers/salgado/salgado.html (accessed February 13, 2009).

Samis, Peter (2008). Who Has the Responsibility For Saying What We See? Mashing up Museum, Artist, and Visitor Voices, On-site and On-line. In Jennifer Trant, David Bearman (eds.). *Museums and the Web 2008: Proceedings*. Toronto: Archives & Museum Informatics. Published 31.3.2008. Available at: http://www.archimuse.com/mw2008/papers/samis/samis.html (accessed February 13, 2009).

Happily Lost in the Virtual Space?

Nico Carpentier

1. Introduction

Nine exhibitions that had run in the Muhka between May 2000 and February 2002, were given a slot on the "Muhka 2000 2001" cd-rom. On this cd-rom, which was self-produced (with the aid of QuickTimeVR), a simulation of the three-dimensional museum space was created. The visitor to the cd-rom thus can move amongst one or more series (of interlinked photographic) representation(s) of the museum. This virtual walk could take him (or her) up to the roof of the Muhka (Figure 2), where he can enjoy the panorama. Or he could view the spaces in their every-day manifestations,[1] from lift doors to fire extinguishers (Figure 3). The basic reason for creating this walk is to display the nine exhibitions, and to enable the visitors to interact with the contents of the museum.[2] The former director of the museum says the following in this regard:

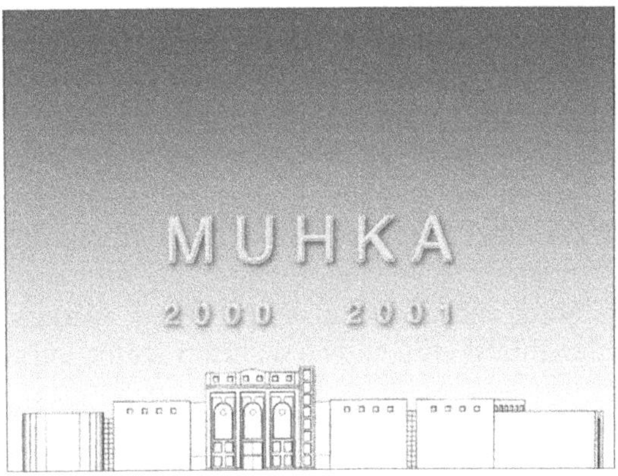

Figure 1: The first page of the cd-rom

> Actually one can more or less move about as in a real exhibition. [...]
> One enters and goes to one exhibit. One feels it, one sees it in the signs coming up whether you should change direction or not. Whether you should click with the little hand, and continue. Thus I found it rather user-friendly. It is merely that, hey.

1 The absence of people in this museum space, which to a great extent colours this mundaneness, will later be discussed.

2 In both meanings of the word: content as meaning, and content as filling of a physical space.

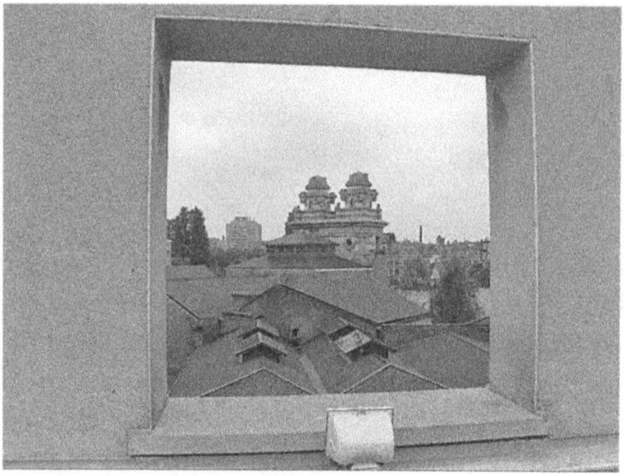

Figure 2: The roof of the Muhka

> Entering an exhibition, going to a second . . . a third hall, a next exhibit. (Translation of an interview with former director of the Muhka.)

This article is an evaluation of the project with regard to access, interaction and participation[3] which the cd-rom offers (or does not offer) the visitor to the museum. From a user-friendly perspective the focus is in particular on the interaction of the user with the interface, on the one hand, and on the other, the interaction of the user with the contents of the cd-rom. To reinforce a personal analysis of the cd-rom, interviews with four people involved in the project[4] are included, as well as the diaries of six students from the VUB.[5] These students all kept a diary during October and November 2002 in which they gave a detailed description on seven consecutive days of their experiences as users of this cd-rom. This small-scale qualitative analysis enables a testing and refinement of the personal analysis, while at the same time explaining and illustrating it with quotes from the interviews and diaries.

3 The AIP-model is used for this. For a more detailed theoretical discussion on this matter, see Carpentier (2011).

4 The following were interviewed: the former director of the Muhka, Flor Bex, the communications officer of the Muhka, Olivier Brems, the producer of the cd-rom, Daniel De Backer, and a member of the desk staff co-responsible for selling of the cd-rom, Lutgarde van Renterghem.

5 The six VUB students were: Jo De Backer, Stefan Hardonk, Andreas Loengarov, Adriaan Moreels, Hanne Van Baelen, David Willems and Femke Wouda. This formed part of their studies in *Theoretische aspecten van de cultuur- en televisiestudies*. Every 'virtual' visit had to last at least an hour, and had to be described in as much detail as possible. At the end of the week they also had to answer a short questionnaire. My sincere thanks to them for their efforts.

2. The objectives of the cd-rom

The objectives of this cd-rom were two-fold. In the first place it was created for internal use: it had to comply with an archival need of the museum. Once again the former Muhka director may be quoted:

> I am […] very happy with the result, because it exactly agrees with my frustration that the exhibition could not be preserved, that it would merely disappear. The artist creates an art work, and the exhibition maker creates an exhibition. And that is lost; nothing is left of it. Thus that was my first objective: to capture and preserve that. This was achieved by means of technology. It originated from a personal need to have something like that, and not with the aim of making a commercial product. (Translation of an interview with former Muhka director)

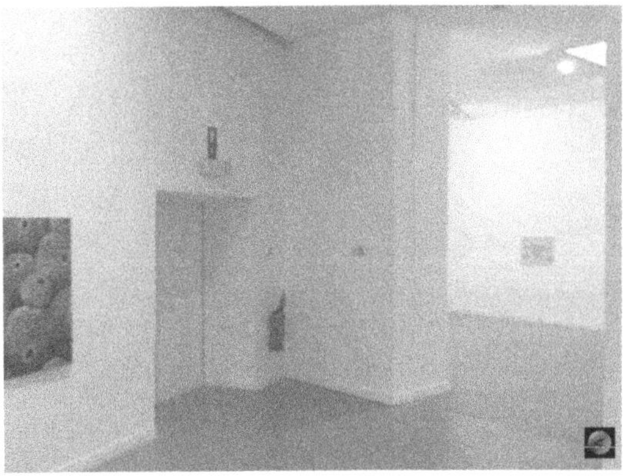

Figure 3: Fire extinguisher in museum

The last part of this statement refers to the second, external, objective of the project, and it contains a communicative as well as a commercial component. The communicative component is closely linked to the two fundamental objectives of museums: to exhibit their collections to a public, and to make the collection accessible for scientific research. Against this background some of the exhibitions put on cd-rom were in fact in the past put on the website that was then used. The communication officer for Muhka, said the following in this regard:

> This fits in with the accessibility of the cultural inheritance […]. I also note that, because of my interest in art, I visit musuem archives or visit a website. When I am looking for a specific painting, I enter that and come to a museum. Because their collection is online, all these things function. Thus one may recruit visitors and comply with scientific accessibility. (Translation of an interview with the communication officer of the Muhka)

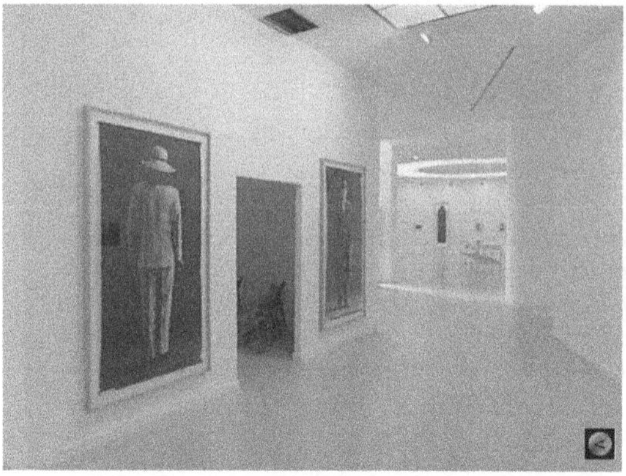

Figure 4: Commencement of James Lee Byars' exhibition

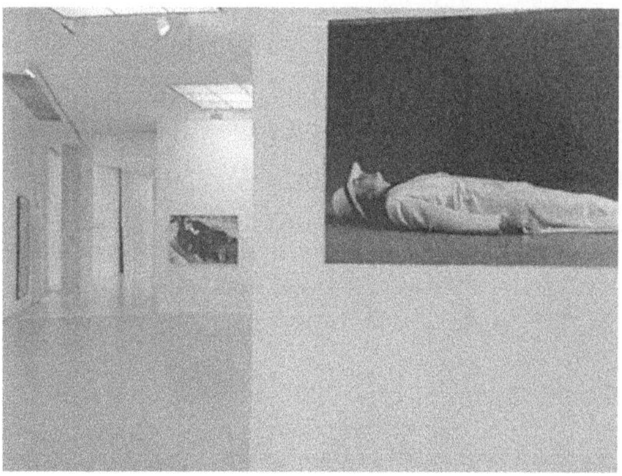

Figure 5: Works by James Lee Byars

As this cd-rom was made in collaboration with an external producer, the commercial aspect also comes to the fore. As Muhka's communication officer said, it in fact was the aim "to not only archive the exhibition, but also to sell the cd-rom to visitors". The aim with the selling was also to partly cover the production costs (which the producer during an interview estimated at 1 million old Belgian francs),[6] as Muhka had not budgeted for this. The producer made the following comment in this regard: "I made the cd-rom with a budget of zero francs, while that of the Louvre was made with ten million." After failure to get sponsoring, it

6 Approximately € 25 000. The production budget of the Louvre (see later) came to € 250 000.

was decided to sell the cd-rom in the museum shop to the Muhka visitors, in order to offer the producer at least some 'return-on-investment':

> At some stage I said, first make a few copies, very cheaply, which we offer for sale at the museum entrance in the bookshop in the museum shop. But as far as I know [...] not very many were sold. [...] The first 100 or 200 copies we repaid to the producer, so that he at least got some money back for all his effort. (Translation of an interview with a former director of the Muhka)

No survey was done amongst the public. However, there were some brief and partly promising contacts with roleplayers from other projects.[7] Sales were however not good. According to a member of the counter staff, only 20 to 30 copies of the cd-rom were sold. The reasons this person gave for this were that it was combination of (1) lack of promotion and selling points, (2) bad timing, (3) vagueness as to the identity of the product, (4) lack of interest in and familiarity with the technology on the part of the visitors, and (5) problems with the demo-version (set up on the counter in the museum shop). These arguments are illustrated below with some comments obtained in interviews:

> The worst problem was merely that it did not sell, and that nothing could be done about it. [...] One should not think that people will discover it of their own accord. Occasionally someone does, but it has to be promoted. (Translation of an interview with producer of cd-rom) (argument 1)

> As we put everything on the cd-rom and offered it for sale at the end of the exhibition year, in January, all the visitors who had already attended the exhibition, and had an interest in it, had already bought catalogues, and they would not also buy the cd-rom with information on all the exhibits. (Translation of an interview with the communication officer of the Muhka) (argument 2 and 3)

> I think that it is still too early for this matter. The funny thing is that almost every household in Belgium has a computer. That it is an instrument that is in fact becoming widely used in Belgium and Flanders. But whether this is too early, well, I don't know. (Translation of an interview with a former director of the Muhka) (argument 4)

> I have always thought that people would rather buy a book through which they could page, than buy a cd-rom. (Translation of an interview with communication officer of Muhka) (argument 5)

---

7 The former director of the Muhka replied thus to the question how they thought the cd-rom would sell: "I had no idea, and also did not keep track. Once I did have a look. "O, a cd-rom of the Louvre in Paris", and then one phones a colleague, and many thousands of them were sold. There was an analysis, and one cannot compete with that, that I know. I thus had few points of comparison with other museums, definitely in Belgium, because no-one in Belgium had ever made something similar." (Translation of an interview with the former director of the Muhka)

> The people who tried it here, found the PC too slow for the cd-rom. (Translation of an interview with a member of the counter staff at Muhka) (argument 6)

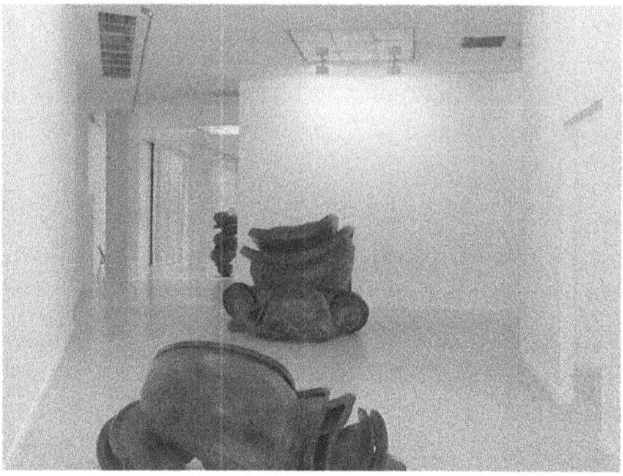

Figure 6: Sculptures by Anthony Cragg

3. What should be done with the cd-rom?

When evaluating this project, one should not only consider the deficient financial and commercial structure, but also the way in which the cd-rom may be used. The project is aimed at a small segment of the cultural-participative scale. But it actually wants to make the cultural products in the museum accessible, whereby, via interaction with technology (computer and cd-rom), interaction with the contents of the cd-rom is made possible. This evaluation is therefore aimed at these specific two components: interaction with technology (and in particular the interface offered by the cd-rom), and interaction with the contents (representations of the museum found on the cd-rom).

3.1. Interaction with the interface

> The user of the cd-rom enters an interface by QuickTimeVR which creates a simulation of the three-dimensional museum space. The producer calls this technology a way in which to represent architecture and art, and related matters. (Interview with cd-rom producer)

This virtual representation of the museum pretends to be a real, quasi-photographic representation of the 'physical' museum. This may be illustrated, inter alia, by the way in which references are made in the diaries of how one navigates and moves through the cd-rom museum. The descriptions used are based on the language used for a 'normal' museum visit:

> I can stroll […] freely through the passageways (diary 1[8] – my emphasis) and the following art work is a depiction of an aeroplane, but it did not hold my attention for long, and I quickly moved on (diary 5 – my emphasis). The visitor quoted above aired the following misgiving on the sixth day: While strolling through the passageways, I had the following thought: it is exactly as in a film. But a film one sees in a hall is different. Beware: it is therefore not better, but the impressions are stronger. This effect one has here as well. (diary 1 – my emphasis)

As the last quote illustrates, we here have a 'different' space than the physical museum space, where other ways of moving and looking are needed. The interface largely determines the characteristics of the virtual space, and the conventions for moving through this space. In addition the producer's choices play an equally important role. When constructing this space, a specific concept was chosen (partly consciously) that was based on discovery. The producer describes it thus:

> I have to say that it was not done 100% intentionally, but we wanted to move away from the over-used things, as are found on the cd-rom of the Louvre. There one sees an arrow, and then everyone knows of course that something else will come. Of course one has to see something, and therefore with us the little hand changes into a little arrow; you do have something. But one gets the feeling, and it is very important, that one is busy with something oneself, thus interactive. One discovers things oneself, and if it is over-familiar, one merely says let's go to page 23, because the painting is there, but then you miss a great deal. It is much more important to discover things. Precisely that is so captivating with interactivity, that you discover things yourself. (Translation of an interview with the producer of the cd-rom)

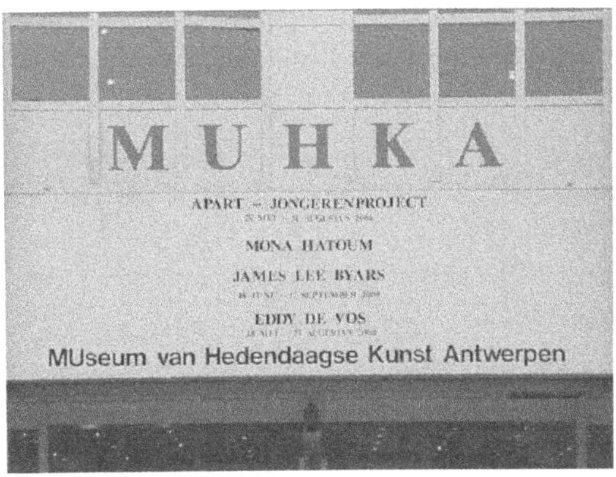

Figure 7: The entrance to the Muhka

---

8 The different diaries were distinguished by means of a numbered code.

The combination of the characteristics of the material space, the strategies of the producer(s) of the cd-rom, and the possibilities offered by the QuickTimeVR technology, largely determine the characteristics of this space. In addition, the users in interaction with the interface, are co-determinants of this space.
That is why the characteristics of this space are described, on the basis of the diaries, in the following terms: confusing, unstable, selective, and partially filled.

Confusing space: "Suddenly I could no longer orientate myself" (diary 1)
The decision to create a cd-rom around the concept of discovery resulted in a combination of discovery and confusion when used. First, the user is confronted with a simulation of a complex three-dimensional museum space, which he/she is viewing from the two-dimensional perspective of the computer monitor. One of the users wrote in this regard: "This confused me somewhat, and it was as if I was wearing blinkers, and had no peripheral sight" (diary 4). In addition the user does not have the required knowledge of the instruments to be able to move through this space, as there are no instructions for use: "You are invited in, and then have to find your own way. There even is no help function" (diary 5). To use Steyart's (2002) words: Because this way of representation (via QuickTimeVR) is so new, the users not only do not have the necessary instrumental and structural skills, but they also do not have the means to acquire these skills quickly. The results of this are well summarised in the following diary entry: "I clicked on everything, but nothing happened" (diary 5). The main lay-out also adds to the confusion. On entering, the users see the main gable of the Muhka on which a list of four exhibitions are shown (Figure 7).

In a number of the diaries the assumption is then also clearly made (not surprisingly) that there are only four exhibitions:

> At that moment it was suddenly somewhat clearer to me when I saw the building and the gable. Via the arrows I came to the front door, and above the door I read that there are four exhibitions. (diary 4)

Later (on the sixth day) this visitor discovered that there are more than four exhibitions on the cd-rom. She also found out how to reach them:

> I have only now discovered that when you first click on the enrtrance portals and then go to the side of the screen, you find thin black arrows. When you click on these, you see other artists than the original four that you see when entering the museum in the conventional way. (diary 4)

Another visitor who also at first thought the one may see only one exhibition on the cd-rom, discovered this, and drily remarks: "I did think that the Muhka was rather small" (diary 1). Accessing the cd-rom is however further hampered by the complex geographical space created by the structure of the virtual visit. Once again metaphors from the 'physical' behaviour comes to the fore: "walks around as if lost" (diary 2), "finds oneself in a labyrinth" (diary 3) or "lost" (diary 3):

> I saw this one room, and wanted to go to the room next to the chair (and the fire-extinguisher). I therefore clicked on the spaces, but suddenly I was totally lost. I kept on reaching a different room trying to reach the original room. […] Suddenly I saw the first room again, and felt relieved. I knew I was not totally lost. (diary 3)

The confrontation with the confusing virtual museum space elicited different reactions. First there is irritation, as demonstrated by the following excerpt from a diary:

> I was becoming irritated because I had no peripheral sight, and had no idea in which direction I was moving. One could only make ridiculously big jumps with the mouse, and I continually went to the wrong side. In addition, if one wants to look closely at something, one has to click on it, and then it comes so close that you have no depth-dimension any longer. (diary 4)

However, it is noteworthy that most diaries mention the learning process that the visitors experience in the first days,[9] and that the irritation is limited. The writers of the diaries have the necessary (general instrumental, structural and strategic) skills to acquire the specific skills needed here. Often the users discover the different functions of the cd-rom by chance, such as turning: "Once again I thought I was stuck, but by chance I suddenly found out how to turn my body in order to see new things" (diary 2). The learning process is not limited to getting to know the interface, but also involves the way in which the complex geographical space is manoeuvred. Orientation remains difficult, but after a while this is no longer experienced in such a negative way, and eventually affords 'spatial pleasure':

> As I became more adept with the instrument, I started enjoying it more. I could study everything unhurriedly, without worrying beforehand which problem I would next experience, and how I would solve it. One can really embark on a journey of discovery, because you do not always know where you will land when you click on something, or what it will look like. In real life you do not have this effect. (diary 2)

Rebellious space: "my mouse lives […] her own life" (diary 4)

It is not always easy to enter the space created by die cd-rom. First, the (PC) users are confronted with a cd-rom containing two portfolios called respectively *muhka2001PC* and *QT 403 PC standalone*. In most cases it needs some scouting

9 It should be mentioned here that keeping the diaries was an instruction to students to use this cd-rom for one hour for seven consecutive days. The question to what degree 'ordinary' users would be willing to cross these barriers, remains unanswered.

around to discover that the first map contains the link to the virtual museum, and the second, the required (QT = QuickTime), which has to be installed if one does not already have a QuickTimeVR. Only once this threshold has been crossed, the software driving the virtual museum may be opened by clicking on the enigmatic *muhkakatabounga. EXE*. Even then further access is not easy, as there are no instructions for use – with the exception of a few sentences on the back of the cd-rom. The two statements below illustrate how difficult access to this rebellious space is:

> It still is not very clear to me what the procedure is to get inside; I just click all over in vain, and suddenly a little arrow lights up, giving me access. (diary 4)

> Getting started is always difficult, because my mouse at first has its own life, and only after much fruitless clicking around, an arrow will light up. (diary 4)

It is not only difficult to enter the museum; leaving certain spaces is not always easy:

> I quickly clicked on the door that I saw, but to no avail. The colour contrast began hurting my eyes, and I wanted to get out. I kept on clicking, but nothing happened; the same objects kept on enlarging, even though I was clicking next to them. (diary 5)

> Suddenly I could no longer exit the exhibition; it really is a maze of passages and rooms, and there is not a single guideline, so that one only walks around, coming to the same rooms time and again. (diary 4)

Unstable space: "The whole thing blocks again. Back to start" (diary 6)

As said above, the basic concept of the virtual museum is to simulate a visit to a 'physical' museum. The visitor is bound to the geography of the 'physical' museum,

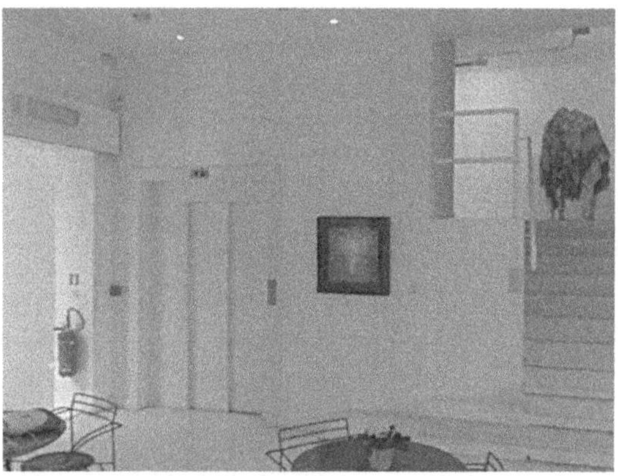

Figure 8: An inaccessible lift

and 'strolls' from one hall to another, in a way which is comparable to a visit to a 'physical' museum. In a number of cases this simulation is interrupted (and so too the strolling), and the reality of the computer again comes to the fore. It is not always only because of 'mistakes' on the side of the users, or of technology breaking down.

Below are a number of examples:

> In front of me another painting appeared. I went to it. Unfortunately it was difficult to make out. It looked as if it was about rocks in a desert. Thereafter I left this hall. Suddenly, by a wrong move, I landed back in the same hall. (diary 1)

> In the end I could find no arrow with which to go forward; I could only go back. But that was an arrow back to the start. I could therefore start the whole thing again. (diary 2)

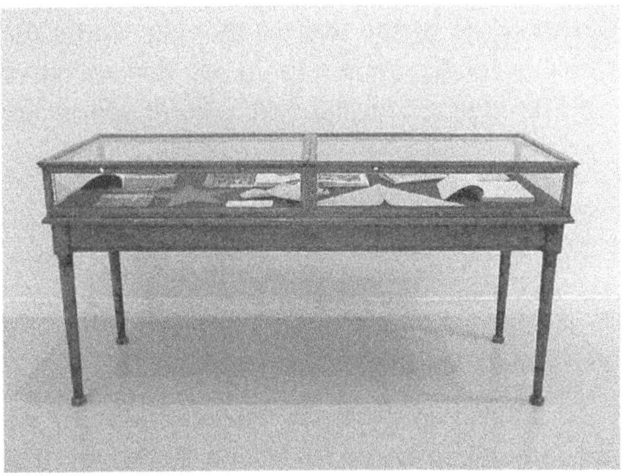

Figure 9: An inaccessible table

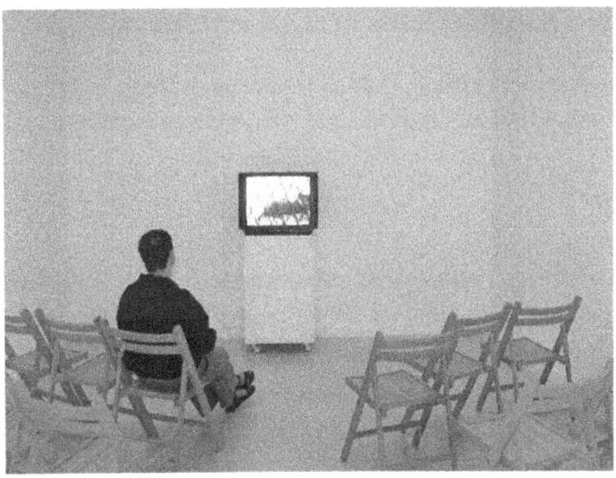

Figure 10: A lone spectator before an unidentifiable video

> I studied them one by one, but unfortunately the cd-rom got stuck. I could go no further. (diary 1)

> There was something: I switched to an mp3-programme, and then the panorama-function of the Muhka got stuck when I returned. (diary 6)

Malfunctioning technology disrupts what McMillan (2002) (inter alia) calls the 'flow': having 'lost' oneself in the virtual environment, there is a sudden (albeit not consciously) experience of a *stopping*, and the user returns to the 'real' environment of the own computer.

Selective space: "Look! You cannot choose" (diary 6)

The interaction within the space created and opened by the interface is simultaneously also limited by the interface. In the first place, it largely directs the actions of the users, and limits the movement space of the user to moving within the space. The users have no other options. Secondly, certain limits are also operative within the concept of moving the self through the virtual space. Some spaces are inaccessible (Figure 8):

> Here there is a lift and stairs on which one may click; try the lift, both up and down? Look! You cannot choose. (db6)

> Let's pass through a mysterious white door which seems to be a service entrance. In vain, you cannot. You can click on it, but then you are catapulted backwards onto the terrace. Magic. Next project: "Magic + Art = Architecture". (diary 6)

Some objects also cannot be viewed in close-up. Both the inaccessibility of the glass tables, where the letters from James Lee Byars to Joseph Beuys are displayed (Figure 9), as the inaccessibility of the audio-visual material (Figure 10) are quoted in the diaries as example. The fact that it is impossible to see the art objects (from every perspective) in detail, also leads to negative reactions:

> The paintings on the walls are once again inacessible. [I thus] cannot study them. Very handy, such a virtual museum, if approximately a third cannot be viewed properly. (diary 4)

3.2. Interaction with the content

This virtual space is not only confusing, rebellious, unstable and selective, but also (and in particular) filled with representations of art objects. The visitors thus also interact with these objects (via the interface). The diaries contain many references to the visitors' appreciation (albeit not positive), and the impact the viewing of these works have on them. For example:

> Suddenly I had to laugh out loud. Eddy Wally had been photographed while lying on the ground. (diary 1)

> The art works that I managed to see were in fact sometimes very harsh, and really moved one. (diary 2)

> On my way outside I thought about the work, and felt somewhat sad about Hatoum's unhappy and lost youth. This is a significant exhibition that will live on. (diary 4)

> On the following picture a number of naked women were apparently having an orgy. Slowly but surely I was getting fed up with all the naked women. And my wish was granted, because then I saw a little monkey in a tree. (diary 5)

The diaries also contained some critical observations on the representation of the art objects. One group of these was about the bad quality of the representations, which was linked partly to a lack of authenticity.

> I found myself thinking that it would indeed be better to visit the museum in real life. There justice could be done to the works. (diary 1)

> However, I really think that more attention has to be paid to the quality of the pictures of the art works. There are various art works of which the representations are very vague. (diary 5)

In addition there is also criticism because certain parts of the museum and of the exhibition that were seen as important were not represented. We have already mentioned the inaccessibility of some of the art objects, in particular all audio-visual works. Adding sound and pictures is one of the (technical) possibilities that the former director of the Muhka mentioned, but these are not present in this cd-rom: "One can insert short extracts of a video, on which one may click and then one can see a bit of the video" (translation of an interview with a former director of the Muhka). The diaries also mention the lack of "a little bit of music […]" (diary 5). Secondly, the fact that the museum is reduced to the exhibited works is noted. There is very little information about the museum itself, but it is in particular the absence of visitors in the simulated museum space (see Figures 10 and 11 for the rare exceptions) that is experienced in the diaries as very strange:

> It would also be much nicer if there were more people present as one strolls through the museum. I did not really feel as if I was in a real museum. (diary 5)

> Two people were sitting there . . . one in a threadbare old coat, and one very hip, à la Nick Balthazar, with closely shaven head and close-fitting t-shirt and black trousers. And a Delhaize-bag on the table, by the figure in red. One cannot zoom in on them. (diary 6 – figure 11)

However, the most important reduction is the absence of explanatory texts. Only when "entering" an exhibition the user gets a whole-screen text about the specific artist. The works do not have any texts. This was a conscious decision, the former director said:

> I wanted to keep it purely visual, and thus I requested the minimum blah-blah; the minimum text. It had to really be a visual experience. [...] And this I oftened said to the public servants, the educational service. If I allowed this, they would have hung texts next to all the paintings. I said no, that is not necessary. Texts you read at home. But in an exhibition you view the works, and don't read any texts. That was my opinion, and it always elicited debate. They still wanted something added. The more text, the better. I said the less text, the better. (Translation of an interview with former director of the Muhka)

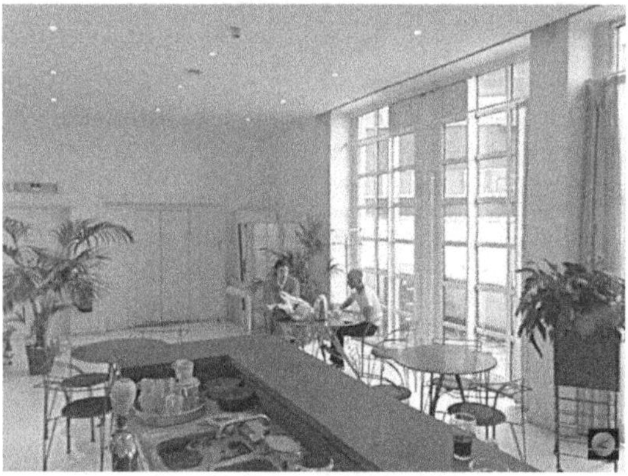

Figure 11: The two people in the cafeteria

The isolation of the contents of the museum also elicited negative reactions and requests for elucidation in the diaries: "I saw paintings that I could hardly understand" (diary 1). Some expressed the wish that the titles of the works should have been given.

> I remembered from the introduction that each painting should have a title, and wondered which title would be suitable for a specific painting. [...] I thought that some paintings are unidentifiable, and it would have been meaningful if we could have been given a title as well as some description. (diary 4)

Others went one step further in their quest for contextualisation: "Often I did not understand the precise meaning behind an object. Elucidation by a guide would have been very useful here" (diary 1). The request for longer explanations was in particular made with regard to one work, namely "Mouli-Julienne (x21)" by Mona Hatoum (Figure 12), which is described in five of the six diaries in great detail. The work is respectively called a "primitive vacuum cleaner" (diary 1), an

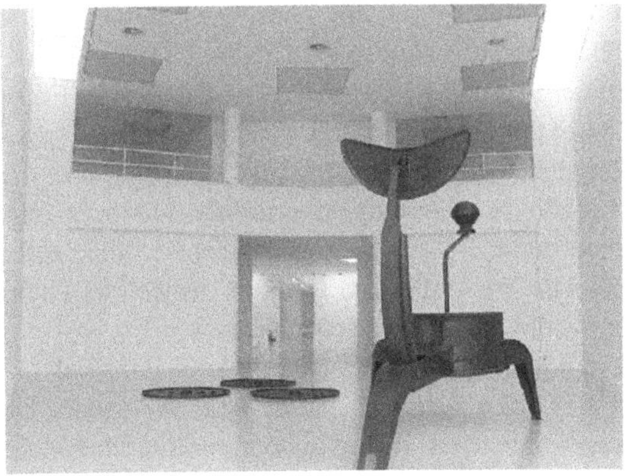

Figure 12: Mouli-Julienne (x21) by Mona Hatoum

"enlargement of something that also exists in real life" (diary 2), a "gigantic egg cutter" (diary 3), a "monstrous vegetable peeler" (db4), and an "object on three legs with three disks in front" (diary 5). Apart from the fascination with the enlarged object, the request for elucidation is particularly important. An illustration of this is also found in the following excerpt:

> In the next hall I found strange objects. Once again I missed an explanation. What was it representing? It looked like a primitive vacuum cleaner, but I was probably wrong. (diary 1)

4. Conclusion

If we consider to what extent visitors may gain access to the cd-rom, and may interact with it, it soon becomes clear that participation in the *production* of the cd-rom (and the inherent meanings) is not possible. The Muhka cd-rom is exclusively aimed at reception, which makes only interaction with the interface and the contents possible. Users of course may gain access to the virtual museum, but only on condition that they visit the 'physical' museum, as the cd-rom and the virtual museum may only be bought there. Only those who bought the cd-rom in effect gets the opportunity to interact with the interface and the museum content stored in the cd-rom.

This Muhka project is problematic in the sense that it is a mix of two approaches: one aimed at the museum, and another aimed at the public. The objective of answering the archival need is indeed legitimate, and it is also realised. However, this focus on the museum aspect has the result that the virtual museum is not public-friendly, and in addition confronts the user with a number of important technological barriers and equally important contextual reductions. These barriers in the first place relate to the confusing, rebellious and unstable space created by the interface itself, and by the usage, both described in detail by die writers of the diaries. Entry was in particular experienced as an extremely frustrating barrier, because expectations in respect of the skills of beginner-users were very high. The fact that the discovery concept was applicable both to the technological interface and to the museum content, caused much confusion in the first phase. Adding more detailed user-instructions than the few sentences on the back of the cd-rom box, as well as a help function and a map of the museum, would make its use much easier.

The diaries show a clear sign of a learning process in respect of interface usage. In some instances they even speak of a definite 'spatial pleasure' caused by 'strolling' through the virtual museum. However, it should be mentioned here that these students had considerable (general) computer skills, and quickly learnt the specific skills needed for manipulating this interface. And the learning process also did not go as quickly as the former director says:

> One merely has to search; move the mouse. […] You know you are on the first page, then the arrow comes, and you click, then there is a circle that says you may turn. Surely you have to master this after a few minutes. It is not so difficult, or what?' (Translation of an interview with a former director of the Muhka)

Thirdly, not all museum spaces could be entered, and not all art objects viewed. The interaction with the geographic museum space thus is far from complete. This spatial limitation is related to the contextual reductions so characteristic of this cd-rom. The diaries mention poor (graphic) quality and authenticity, and the reduction of the museum to its artistic content. In this way both the museum as institution and the museum public are excluded. The most important reduction – resulting from a primordial choice in the production of the cd-rom – came about through isolation of the museum content from secondary meanings. There is almost no interpretation of the art works or the artists. The decision to let the art works 'speak for themselves' and to let the users produce all meaning on the basis of the 'unadulterated' works, elicited negative comments in the diaries. The diaries revealed in particular an acute need for contextualisation, and the desire not to be placed in the position of hyperactive users.

However, the reductions did not prevent the diary-writers-users from actively enjoying the virtual museum visit, and attaching their own meanings to the art works. In spite of all the criticism, the diaries also revealed the pleasure of and fascination with a virtual museum visit – as simulation of the 'physical' museum. The pessimistic connotations that the French sociologist Baudrillard (1983) attaches to the concept of simulation, are hereby softened. More is hidden behind the simulation than stimulation and hyper-reality. Despite the lack of ultimate authenticity in the art works that are represented, and which indeed fits into Baudrillard's critical analysis, the pleasure which the virtual visit to the museum gives, is real.

Acknowledgments

This study was made possible with the support of the Vlaamse Gemeenschap: Programma Steunpunten voor Beleidsrelevant Onderzoek. However, the viewpoint given in the text is the author's, and not that of the Vlaamse Gemeenschap. The Vlaamse Gemeenschap may not be held responsible for the use of the facts offered. An earlier version (in Dutch) of this article has been published in Carpentier (2005). The publishing of this article was supported by the Estonian Science Foundation grant no 8006.

References

Baudrillard, Jean (1983). *Simulations.* New York: Semiotext[e].

Carpentier, Nico (2005). Vrolijk virtueel verloren lopen in het Muhka? *Freespace Nieuwzuid,* 5 (17): 86–103.

Carpentier, Nico (2011). *Media and Participation. A Site of Ideological-Democratic Struggle.* Bristol: Intellect.

McMillan, Sally (2002). Exploring Models of Interactivity from Multiple Research Traditions: Users, Documents and Systems. In Leah A. Lievrouw, Sonia Livingstone (eds.). *The Handbook of the New Media. The Social Shaping and Consequences of ICTs*. London: Sage, pp. 163–182.

Steyaert, Jan (2002). Inequality and the Digital Divide: Myths and Realities. In Steven F. Hick, John G. McNutt (eds.). *Advocacy, Activism, and the Internet. Community Organization and Social Policy*. New York: Lyceum Press, pp. 199–211.

Increasing the Usability of the Museum: Four Studies

Pille Pruulmann-Vengerfeldt

Pille Runnel

1. Introduction

This article presents some of the work done in two grant projects – "Developing Museum Communication in the 21st Century", and "The Problems of Transformation and Reception of Cultural Heritage in the Digital Age", which have both tried to look at the changing relations between audiences and heritage institutions. We will use examples from the Estonian National Museum (ENM) as the basis of our analysis. In this article we will support our argument with empirical case studies and data collected using several methods at different points in time. Rather than being a project report, we hope this will be a more reflexive overview of audience relations in museum institutions and participation. In using the notions of audiences and users interchangeably, we assume that audiences are active despite the communication channels used. In increasing use of ICT solutions in heritage institutions, audiences are placed in even more active roles as users or produsers, as used by Bruns (2006). Benefiting from some of those active users as content providers and active participants in the museum context is a challenge. Coming from a media studies perspective, we approach the publics in the heritage institutions from this angle. In addition, when bringing in notions of usability as conceptualised by Jakob Nielsen in order to add ICT and marketing related knowledge, we try to focus this paper's analysis around understanding how improved usability can support user generated content and participation in museums. By doing this, we try to place this article on the crossroads of different disciplines, as these add value to the practical applications we are working on.

We argue that in order to activate participation in heritage institutions, publics should not be left to find their way around, but rather heritage institutions need to make a conscious attempt to create spaces of participation that are user friendly and engaging.

2. Memory institution choices for user generated content

Participation and active audiences are not entirely new phenomena in the context of heritage institutions (see, for instance, Carpentier, 2010). Many museums have built their collections using objects and information collected from the people.

Historically our home institution, the Estonian National Museum, has regarded the public not only as audiences of their exhibitions and customers of services and consumer products, but individuals and groups have also been approached as subjects of study and as information providers and knowledge co-creators for the collections (related to the research approach of ethnology and folkloristics). The ENM is largely based on collections and archives about folk culture, communities, everyday life. A remarkable archive has been formed on the basis of written contributions and donations from the museum's network of correspondents (formed more than 70 years ago). However, it can be said that in all these participatory activities, the heritage worker has played the role of the gatekeeper, moderating and limiting the participation for particular purposes. The ENM, similarly to other museums, has been the legitimate producer and guardian of heritage and social memory and through these roles the museum has also been helper and teacher in developing a sense of collective identity and citizenship.

Today, online environments seem to provide ample opportunities for the public to engage in dialogue with heritage institutions. However, two-way communication assumes not only the existence of the communication channel, but also willing parties who are interested in communication. Despite the promising online opportunities, the continuously hierarchical and traditionalist nature of heritage institutions may be part of why audiences/users are not that keen on participating and contributing. In particular the sense of expert power or the consideration of an abstract 'them' who know better than the 'average man' can become an obstacle to participation, even in the time of these potentially democracy-facilitating technologies. Thus the obstacle of perceived lack of expertise has to be overcome on both sides. All participating parties need to understand that the 'average man' is an expert in his own life and his contribution can be valuable for the museum, even when they are less than perfect by the museum's standard.

When we conducted the interviews with cultural heritage institution workers they summarised their passive attitude towards creating online content by claiming that they are not experts to comment or to say, and that their opinion doesn't matter and nobody cares (Aljas, Pruulmann-Vengerfeldt, 2009). In a way, museum workers 'lost' their expertise when facing online environments and became hesitant in providing their contributions. This also indicates the perceived sense of expertise when considering participation from the general public in their field of expertise. In the past, gathering input from the society and mediating information were tools for the memory institutions to maintain an expert position. That position can now, when technology permits potentially very open and wide participation (and content creation) at low cost, become threatened.

> The idea of becoming virtual might not be a pleasant one for some museums, especially not for art museums who cherish the ideal of the 'real thing' and its aura. However, this development is inevitable because of the increasing digitisation of cultural heritage and the demand to make collections more accessible. Eventually, these trends will blur the differences between cultural heritage institutions, and in the long run these institutions will merge into one memory institution. (Schweibenz, 2004).

These interlinked memory institutions hailed by Schweibenz above, show that the important user motivation comes from the content itself. National museums and ethnographic collections in particular can claim to be 'living' museums where communities connected with that particular museum can add to the collection as they experience everyday life. There have been arguments (Carnall, 2009) which say that memory institutions have had significant obstacles that have stopped them from being online to any great extent. These obstacles have included the genuine fear that people would stop coming to museums if they could access museum collections online (ibid.).

At the same time, the virtual museum can extend the ideas and concepts of collections into the digital space and in this way reveal the essential nature of the museum.

We have now briefly mapped different kinds of obstacles and considerations related to the changing roles of the museums and museum audiences in relation to the new technologies and user-generated content online: museums being hesitant to lose control; audiences as potential participators facing the still-present barrier of professional expertise; experts being afraid to lose visitors entirely because the content of their collections has been made available and open. Now we will turn to the case study of the Estonian National Museum, which deals with how the museum faces these threats and finds motivation to use online and on-site technologies to support participation by the general public.

3. Participatory actions at the Estonian National Museum

As mentioned before, the ENM has a long-term tradition of collecting tangible and intangible heritage. Along with the spread of internet technologies, online communication and collection have required conceptualisation and rethinking of collecting in general.

Relying mainly on low-tech applications or interactive screens in exhibition rooms, the ENM has also expanded its presence to Second Life, Twitter and Facebook for the sake of dialogue with its audiences. Online and offline[1] participatory options at the ENM are more geared towards facilitating dialogue with users –

1 Many of the offline stories were created with computers – cd-roms with digital images, word-processed texts etc., but as they were delivered using regular postal services or brought to the museum by hand, these are counted as offline contributions.

asking them to comment on and to add to digital collections as a complex body of knowledge. In the context of this article, we will not discuss the effectiveness of those online social networking tools for the museum.

We will focus our analysis on four different examples of activities, which combine both online and offline and which have been designed to facilitate user participation and possibly the breaking of the barrier between museum experts and museum audiences on the level of both collections and exhibitions (Table 1).

We will analyse participatory actions conducted at the Estonian National Museums in recent years from Jakob Nielsen's usability perspective, starting firstly with a short introduction of these examples. We have summarised the examples following the inspiration of Mariana Salgado (2009) into an overview table, illustrating the initiatives in chronological order and summarising the online on-site dimensions of each of these initiatives. Our own position in analysing these initiatives is that of auto-ethnography and production ethnography as we are initiators of these initiatives as researchers and as museum workers.

Table 1: Overview of the participatory activities combining online and offline

		"Give the Museum a Day of Your Life" (April 2009)	**Exhibition comments with pen and paper (autumn/ winter 2010)**	**Museum night exhibition (May 2010)**	**Open Curatorship exhibtion (2010–2011)**
Aim	Overall aim	Collect stories of everyday life of Estonians on 14th of April	Receive comments and clarifications on photo collection displayed in exhibition	Receive stories and comments on permanent exhibition	Hosted exhibitions invitation to general public to display their own collections or cooperation
	Museum activity	Collecting	Displaying/ collecting	Displaying	Partnership

Access point	Online	Advertisement online, e-mail address and online form as submission sites	Advertisement of exhibition, no participation possible	None	Online submission of exhibiton proposals, Online evlauation of exhibition, proposals by public vote
	On site	Postal and hand-delivered stories	Post-it notes stuck to favourite pictures worth commenting on	A5 papers on washing line and pegs for commenting on specific aspects of the exhibition	Offline submission of exhibition proposals. Offline display and evaluation of proposals. Two winning exhibitions displayed at the main museum building
No of participants	Online	23	NA	NA	28 online proposals 509 online votes
	On Site	402	80	17	5 on-site proposals 55 on-site votes

3.1. Collecting – "Give the Museum a Day of Your Life"

At 2009, the Estonian National Museum carried out a campaign called "Give the Museum a Day of Your Life", which took place both online and offline, aiming to document everyday life in 2009 and asking people living in Estonia to contribute. People were asked to document the April 14, 2009, which was the hundredth anniversary of the museum. The contributions were included to the collections of the ENM. Contributions to this kind of initiatives need dedication and time from the people. They know that the stories and photos or other visuals became part of the museum's collections and archives.

This adds a sense of value and motivation to the audiences to participate. At the same time, the topic remained simple enough for everyone, as all of us can claim to be an expert in one's own personal everyday life and everyday practices and thus the barrier to participation was low.

3.2. Exhibition comments with pen and paper

At the same year, the ENM held a photographic exhibition called "With a Thousand Steps", based on an overview of the museum photographic collections and on a display of every 1000th photograph from its collections. Here the user-generated content experiment was carried out. Visitor participation was made easy – people were given the opportunity to add free-form comments on the presented photos by using post-it notes and pens at the exhibition hall. Motivation for this experiment was provided through a prize draw where participants could expect to receive a particular photograph as a print for their personal use. Eighty percent on the comments contributed were expressions of emotion, such as 'beautiful', 'great', 'I like', and/or 'I would like to have that too'. However, from the heritage institution's point of view, the more valuable were the remarks that indicated the new knowledge that people got from photos or where exhibited photographs activated new interests or questions. In addition, a few corrections were made to the photograph legends the museum had: "It should be Artur Vasiksaat, because the name Vasikraat doesn't exist in Muhu" (Example from the post-it notes on the photographic exhibition).

This kind of 'tagging' experiment indicates that in order to have user-generated content, there is not necessarily a need for high-tech solutions and expensive software.

In 2010, the ENM also used pen-and-paper technologies at Museum Night, at which people were invited to tell their stories about the permanent exhibition (or parts of it) of the ENM. In 4–5 hours, 17 stories were collected, ranging from short exclamations to longer personal recollections.

3.3. Open Curatorship exhibition

In order to promote the idea of open curatorship, the ENM called audiences to propose exhibition ideas, giving the winners the opportunity to develop a real exhibition. The open curatorship project was thus participatory on three levels: proposing the ideas, selecting the winner and developing the exhibitions was all done by people who, under usual circumstances, might be the more or less passive visitors to the museum. In order to facilitate proposing the ideas, a set of questions was formed for prospective participants.

Altogether 33 ideas were proposed and 569 voters participated in selecting the winning project. The winners were selected in two categories – exhibitions based on artefacts from outside the ENM and exhibitions based on the museum's collections. Although the museum workers expected to have some unusual approaches or choices of topic, of the sort that do not usually make their way to the exhibition halls, the reality was different: the winning project, based on material coming from outside the museum, was dedicated to the changing funeral traditions in Estonia and the winning project using the museum's own collections displays copies of ethnic dress and folk art produced by a handicraft cooperation called Uku.

People were asked feedback to how this kind of participation in the museum's activities is similar or different to their usual engagement with the museum.

4. Increasing user motivation for participation

In his analysis, Nielsen (2006) argues that one cannot change the principle that the majority of the users of a particular online environment will remain passive. However, he argues that there are five key possibilities to have users participating in content creation.

Nielsen (2006) argues that these are: (1) making the participation easier, (2) providing the possibility to edit rather than create, (3) promoting quality contributors, (4) making participation a side-effect of visiting the site and (5) rewarding, but not over-rewarding, participants; by doing so one can engage more than the 90:9:1 ratio mentioned above.

This provides us with additional clues as to what user motivation might be for participating with the heritage institutions (online) and we are going to discuss the experiments of participation from the usability perspective provided by Nielsen to help understand how these ideas can be put in practice from the heritage institution's perspective.

4.1. Making participation easier

Nielsen's (2006) first key proposal is making participation as easy as possible, which indicates that the usability aspect of the participation environment will be important. By showing people that contributing is easy, one might end up with contributions that might be less valuable for the heritage institution but would help in making participation a habit. Here the offline example of tagging a photographic exhibition with paper post-it notes is an excellent example. Using low-tech solutions and familiar technologies (paper and pen) the participants did not need to learn new practices and could participate while viewing the exhibition. However, experience from the same exhibition indicates that when participation is made too easy, this could also reduce the quality of the contributions. 80% of post-its provided at the exhibitions focused just on utterances like "beautiful" or "I'd like to have this" alongside the rather carefully considered quality contributions. These posts activated the user, but they do not necessarily support the creation of quality contributions relevant for the museum.

Also an open curatorship exhibition introduced public voting as an easy method to participate in museum activities. We made online and paper-based voting questionnaires on which everyone could give marks for either all exhibitions or just highlight their favourites. This form of participation was popular, resulting of 569 people casting their votes.

4.2. Providing the possibility to edit rather than create

Secondly, the notion that editing is easier than creating (Nielsen, 2006) was taken into consideration when preparing the "Give the Museum a Day of Your Life" action. Before the public call for participation museum staff made different attempts to collect their own lives and these stories were provided as examples of different styles of participatory content to help people overcome the complications of starting. The idea that editing existing content is easier than starting from scratch is also visible in Trant (2009). The idea supports Carnall's (2009) observation that online museums are typically very content-light, which makes it difficult to attach the contributions of the public to specific pre-existing structures. Contribution to the online content has been related to the network effect (Liebowitz, Margolis, 1998), meaning that the resource becomes more valuable when there are other people consuming the same good. Thus internet users expect and like to contribute where others are and where some prior content already exists. The more information and opportunities to link, add, comment on and tag the information heritage institutions have online, the more valuable the resource is for individual users. Hence the heritage institutions, like museums, which have so far been very shy of adding their digital information online in the fear of losing their real visitors (Schweibenz, 2004) have to overcome that fear in order to make users interested in generating content on their sites.

Nielsen's second key statement – that editing is easier – was also visible in the open curatorship exhibition. For the exhibition proposal, a set of questions was formed in order to encourage people to start thinking about an exhibition idea. Of the 33 ideas submitted, the majority follows the proposed pattern, although a number of others resisted the provided form and looked for other ways of giving their own ideas. Those who choose to ignore the set questions usually had some previously formed specific ideas and found it difficult to suppress or extend those pre-existing ideas into an exhibition proposal. Thus, when providing editing or clearly pre-formatted ways of participation, it is worthwhile to consider other options for those for whom such pre-formatting might be too limiting.

Finding ways to support participation should avoid generating new barriers. The Museum Night pen-and-paper comments experience shows that exhibition items themselves can become the templates for further editing. Our participants used the things they saw as the basis for their stories, connecting objects with their own experiences.

These stories made most sense when seen directly next to the relevant parts of the exhibition. In future, sourcing and displaying such stories may become an inherent part of the new permanent exhibition, while in the context of this particular Museum Night the stories became additional focal points of the exhibition.

4.3. Promoting quality contributors

Thirdly, Nielsen's idea (2006) that promoting quality contributors would foster participation is related to what Marlow, Naaman, boyd and Davis (2006) indicate as sociable interests and motivations of the users. Marlow et al. (2006) lists the following aspects as the social motivations for tagging: communicating and sharing, attracting attention, self-presentation and opinion expression. However, they point out that users might not always consider the social (or even altruistic) motivations as the most important ones. People often start for personal organisational reasons and later move on to the social benefits (ibid.). In addition, digital collections, which make user contributions official parts of the heritage institution's collections, are a good way of promoting participation. This was also part of the "Give the Museum a Day of Your Life" initiative as potential contributors were assured that their 'days' will be part of the museum collections.

The open curatorship exhibition was a participatory action in which the promotion was inherently part of the action. Each of the exhibition proposals was displayed on the website and also on a temporary stand in the museum foyer. The winning exhibitions were selected by the voters on the website and in the museum foyer and winners also received the honour and responsibility of actually displaying the exhibition.

4.4. Making participation a side-effect

The fourth aspect of participation is more of a technical feature (Nielsen, 2006) in which the system itself points out that if the user found a certain kind of content relevant, they might also find other information helpful. This feature is not used as part of usergenerated content (preference data sourced from users) but is used as an expert-provided linking of the materials in which one digital item can be linked to another through an expert-provided relationship. This way of linking the collections has been used in the Estonian Literary Museum online project, Kreutzwald's Century[2] in which the user can explore history in a non-linear way, thus creating (though not leavin, g a record of) her or his own trail through literary history. Here one potential application of user-generated content is to store the trails of the digital content users and provide them as potential pathways to

2 Kreutzwald's Century (http://kreutzwald.kirmus.ee/) is a web portal of Estonian cultural history, developed in the Estonian Literary Museum, which introduces the cultural history from the begginning o fthe 19th century – 1918, a year, when the Republic of Estonia was established. Its content is primarily the digital archive: archive materials, photos, earlier literatuure in Estonian language, introductions of the figuures in Estonian cultural history in the context of historical events of the time.

those interested in non-linear narration of the literary history. Here, a great future potential also lies in recommendation systems supporting participatory activities, which do go beyond the model of commercial providers (Amazon) of outlining similarities between products. In the Estonian National Museum, the new permanent exhibition, currently being developed, looks for ways of providing a social recommendation system highlighting relevant marked units from the collections or additions from the users, based upon the material in which a user has already expressed an interest. Such social recommendation systems also have the potential to provide recommendations based on both familiarity/similarity while at the same time enabling conscious comparison and connection of objects, stories or comments that perhaps represent a different kind of perspective and use other people's recommendations to help in the connections.

4.5. Rewarding, but not over-rewarding participants

The fifth consideration for increasing user motivation is rewarding active participation. Here the reward can be monetary or giving away things and this way of rewarding has been used at the Estonian National Museum, where a prize draw was included as motivation for both of the discussed user-generated content initiatives. However, it did have some drawbacks as the promised prize for the post-it note participation involved a print-out of the chosen photo; the many comments were mainly related to the beauty of the photo. Rewarding quality contributions can also potentially invite concerns about how a judge or jury decides on the 'best' contribution. This might also have drawbacks when considering that there are groups of people who are less likely to contribute as they might not perceive their contribution to be 'worthy' or 'good enough' for the competition.

Furthermore, there are ethical implications from the perspective of research if the contributions are intended to be used in scholarly work about a particular topic in which one expects to collect material and avoid impact from the researcher's side (forcing expert power through giving awards) as much as possible.

Another example of rewarding participants was the open curatorship project. Here the reward was the opportunity to execute the exhibition proposal later on in museum space. In many ways the reward was intended to be selection criteria for participant – to find those willing and interested in realising their idea. But in this way the reward also worked as a possible barrier.

4.6. Summarising the initiatives

In summary of the initiatives above, we have drawn Table 2. As it is possible to see, the "Give the Museum a Day of Your Life" and Open Curatorship Exhibition initiatives have managed to take into account the variety of Nielsen's recommendations. Here the online and on-site combinations have worked well to provide alternatives and support participation in variety of ways.

Table 2: How online and on-site initiatives provide support for participation

	"Give the Museum a Day of Your Life" (April 2009)	**Exhibition comments with pen and paper (autumn/winter 2010)**	**Museum night Exhibition tagging (May 2010)**	**Open Curatorship Exhibition (2010–2011)**
Make easier	Providing different formats	Familiar technologies, available at multiple points throughout the exhibition		Providing different formats
Edit rather than create	Providing a variety of different samples	None	Providing questions. Providing pre-filled papers	Providing structure for the proposals. Proposing survey templates for voting
Promoting quality	Selection of the best stories, but no public promotion of those	None specifically, all contributions were visible for all visitors	None specifically, all contributions were visible for all visitors	Winners rewarded with exhibition
Side-effect	Not available, although participation becomes an added value for others			
Reward	Prize draw general	Price draw related to the exhibition	None	Exhibition as a reward put some people off

Although Nielsen's (2006) framework was initially designed to support people attempting to build online communities, it also supports analysis of the participation initiatives for the practitioners, who can then look at the strengths or weaknesses of their activities. While Nielsen talks mainly about online issues and usability in general, the ideas proposed, and the support provided, suit on-site activities because this makes some things simpler. Here in our examples, we have managed to implement many of the recommendations, especially with online initiatives. However, the framework also indicates that there are issues that could have been implemented better and profiles of the participation project could have become more visible, especially if quality contributions could have been better promoted.

5. Conclusion

The ways museums use technical measures to support or facilitate participation have to be linked with what museums have set as an aim for that particular participatory action.

When participant numbers are sought after, ease of participation is very important, but at the same time, this ease can also become a barrier to more complex or different contributions.

No matter if the participation is high-tech or low-tech, technology should not become a barrier in itself. In today's world this means replicating participatory initiatives online and offline, because some technologies will become more accessible to some groups than to others. In our experience, intertwining works best for the engagement of a diversity of groups.

When outlining the different ways in which museums can engage audiences, Simon (2010) stresses the importance of the 'why' of participation. Chosen technological means should support the overall aim. Participation should not be an aim in itself, but rather support the other goals and activities of the museum.

In conclusion to the particular participatory actions, it could be said that neither of the used participatory activities 'threatened' neither the museum nor the audiences/users on the level of expertise and knowledge making. Rather, the audiences received a different kind of approach to museum collections and exhibitions, and to some extent, other visitors. The actions also added reflexivity to museum professionals' activities. In our examples, the contributions collected through these different measures have raised debates around issues of quality and issues of validity. On the one hand, museum workers consider 'average man' to be an expert in his or her own life, while on the other hand they are critical and hesitant when it comes to the idea that everyone should record their lives, provide input to the museum or support the collections with their own stories. For them the question of validity, standards of quality and the value of the contribution needs to be discussed and possibly re-evaluated in the light of increasing participatory activities.

Combining online and offline technologies is crucial for future museum institutions. In the future of the Estonian National Museum, the variety of participatory activities both online and on site will hopefully help to form a strong network of people around the museum. We expect that in the future technology will support museums and audiences, for example through re-making the 'museum visit' into the 'museum engagement': with content-rich ICT solutions, open to participation, with which a potential museum visit would start long before the real visiting experience and would not end with leaving the museum. Participatory contributions would be seen as an integral part of the museum and discussion around their value would be open with the participation.

Acknowledgements

The publishing of this article was supported by the Estonian Science Foundation grants no 7162 and no 8006, and the Estonian Ministry of Education and Science target financing projects no SF01800002s07 and no SF0180017s07.

References

Bruns, Axel (2006). Towards Produsage: Futures for User-led Content Production. In Fay Sudweeks, Herbert Hrachovec, Charles Ess (eds.). *Cultural Attitudes Towards Technology and Communication 2006*. Murdoch: Murdoch University, pp. 275–284.

Carnall, Mark (2009). Chasing the Online Audiences. *University Museums and Collections Journal* 2: 37–39, available at http://edoc.hu-berlin.de/umacj/2/carnall-mark-37/PDF/carnall.pdf (accessed February 19, 2011).

Carpentier, Nico (2010). A Short History of Cultural Participation. Agnes Aljas et al. (eds.). *Transforming Culture in the Digital Age: International Conference in Tartu 14–16 April 2010: Papers Presented at the Conference in Tartu, 14–16 April 2010.* Tartu: Estonian National Museum, Estonian Literary Museum, University of Tartu, pp. 11–19.

Liebowitz, S. J.; S. E. Margolis (1998). Network Externalities (Effects). In *The New Palgrave's Dictionary of Economics and the Law*. MacMillan, available at http://www.utdallas.edu/~liebowit/palgrave/network.html (accessed February 19, 2011).

Marlow, Cameron; Mor Naaman, danah boyd, Marc Davis (2006). *Position Paper, Tagging, Taxonomy, Flickr, Article, ToRead.* World Wide Web 2006 (WWW 2006): Collaborative Web Tagging Workshop, Edinburgh, Scotland, available at http://www.ibiblio.org/www_tagging/2006/29.pdf (accessed February 19, 2011).

Nielsen, Jakob (2006). *Participation Inequality: Encouraging More Users to Contribute*. Jakob Nielsen's Alertbox, available at http://www.useit.com/alertbox/participation_inequality.html (accessed February 19, 2011).

Pruulmann-Vengerfeldt, Pille; Agnes Aljas (2009). Digital Cultural Heritage – Challenging Museums, Archives and Users. *Journal of Ethnology and Folkloristics* 3 (1): 109–127.

Salgado, Mariana (2009). *Designing for An Open Museum. An Exploration of Content Creation and Sharing through Interactive Pieces*, available at https://www.taik.fi/kirjakauppa/product_info.php?cPath=23&products_id=157 (accessed February 19, 2011).

Schweibenz, Werner (2004). Development of Virtual Museums. *ICOM News* 3, available at http://cool.conservation-us.org/icom/pdf/E_news2004/p3_2004-3.pdf (accessed February 19, 2011).

Simon, Nina (2010). *The Participatory Museum.* Santa Cruz (Calif.): Museum 2.0.

Trant, Jennifer (2009). Tagging, Folksonomy and Art Museums: Early Experiments and Ongoing Research. *Journal of Digital Information* 10 (1), available at http://journals.tdl.org/jodi/issue/view/65 (accessed February 19, 2011).

Acknowledgements

The authoring of this article was supported by the Estonian Science Foundation grant no. 7816 (...) and the Estonian Ministry of Education and Science target financed projects no. SF0180002s07 and no SF0180001s07.

[illegible]

CONCLUSION

Visitors, Users, Audiences: Conceptualising People in the Museum

Pille Runnel

Krista Lepik

Pille Pruulmann-Vengerfeldt

1. Introduction

The modern age is characterised by profound changes in society, including weakening social ties and the changing relationship between citizens and states, as well as the changing position and role of institutions in the social system and social order (Giddens, 1991; Sztompka, 1999; 2003). Cultural institutions and museums in particular can be seen as institutions that have been part of the institutional fabric of a society, often articulating and representing dominant views about state, nationhood, education and people's roles and responsibilities in any given society. Today, this position has become more ambivalent, as museums carry the potential to fulfil new tasks, including challenging dominant hegemonic culture and providing alternative views about society (Mason, 2007).

While museums are established to present the nation, professionals working in them may not consider their institutions to be national in a political sense but rather as being a national service (National Museums, 2012: 14), offering neutrality, objectivity and trust by applying different representational strategies for presenting the past. Museum curators, responsible for telling stories about the past, rely on two opposing approaches. Firstly, there is the understanding of culture that looks at culture as largely static, objective and shared – something 'real'. Secondly there is a more liberal, scientific approach, theoretically grounded in new museology, that is open to interpretations of the past (Runnel, Tatsi, Pruulmann-Vengerfeldt, 2010), inviting curators (researchers, experts) to take up a dialogue with visitors, and which understands museums as sites of cultural participation.

Struggling with these opposing positions, museums need to pay more attention to how they define different stakeholders and groups in and around the museum. Traditionally the museum institution has been influential in terms of holding power to produce knowledge and accepted truths for museum visitors, positioning the visitors as passive recipients of cultural content curated by museum professionals. McLean (1994) has observed that today, "from a situation where the public had little say in museum affairs" the relationship between museums and their audiences has changed to "one where the sense of the public is an overriding factor" (McLean, 1994: 244). Behind this change, two important trends can be identified: first, democratisation and cultural participation is becoming part of the museum field, and secondly, museums are becoming competitors in the field of leisure time services.

Taking up the first aforementioned trend, the transformation of societal context (the dramatic rise of technology, including participatory media technologies and related transformations of individual lifestyles) has democratised the contemporary museum. Here, the change can be theoretically embraced by the development of the participation concept that emerged in the political sphere, from where it spread to different areas, such as fields of social and cultural participation. The transformation of society and technological development allowed discussion of participation in other fields (Carpentier, 2011), including museums. Indeed, museum studies claim that museum visitors as an active and heterogeneous public (Macdonald, 2006) have gained more options to interpret information and to interact with the museum in a more reflective manner. Visitors' capabilities to participate in culture, and their trust and openness towards contemporary museums are important in order to comprehend the consequential museum role in fostering participation and supporting democratization of the society at large.

Addressing the second point, contemporary museums as actors in the field of leisure services has to be mentioned. Today, most museums exist in order to attract and serve visitors (Falk, 2009). By competing with many other organisations and institutions for the public's leisure time (ibid.: 186), museums are forced to market themselves to diverse consumers. This disposition of wanting to reach the public increases not only the heterogeneity of audiences but also reinforces the role of the welcomed audience.

The changing role of the audience accentuates a museum visit as a meaningful knowledge construction experience rather than a controlled act of passive receiving of provided information. At the same time, while society is changing and museums work hard to change their relations to their visitors, large segments of the populace remain uncritical towards the information presented in museums and the ways it has been produced. Yet high public trust of museums fails to recognise that scholarship, interpretation, and controversy are central to what museums do and therefore the public needs to be made aware of the context in which museums are created and in which they operate (Scorrano, 2012: 347–348). Uncritical attitudes are observed in museum visitors towards information, representations and interpretations disseminated by the museum institution. This places museums in a difficult role and shapes the active audience concept. Commenting on the process of identity construction, Scott (2007) calls visitors "active ideological agents" projecting their own complex meanings onto exhibitions, although with limitations of a concrete cultural framework to the universe of meanings they make. Visitors interpret museum exhibitions through their own personal experiences, histories and beliefs (Crooke, 2007: 107) as well as on the basis of certain narratives that predominate in wider culture and thus also in the museum visitor's perception (Scott, 2007: 2). This acknowledgement makes us question how to best understand and define the visitors or museum audiences and how this relationship is forged.

In the next section we will look at the ways in which people can be conceptualised on the basis of their varying relationships with museums. We will also look at what museums as institutions can do to foster active audience engagement and meaning making. Finally the article looks at the ways in which museum audiences themselves represent certain properties as preconditions of becoming attentive, critical and engaged participants.

2. From public to participants

The interdisciplinary nature of our research team is reflected by what some would call discursive confusion in this book. We take turns in referring to the museological concept of 'visitors', the notion of 'users' originally emerging from internet studies, we use the concept of 'participants' similarly to how it has been used in the political sciences and introduce the vaguest of all, the notion of 'audiences' originating from media studies. We have brought these notions into our analysis as a result of long-term research where we came to see these terms as interrelated, and, to a certain extent, as replacing each other. It is important to point out that these concepts can also been seen as a hierarchy of relationships that people have with the museum. This hierarchy, outlined as a ladder in Figure 1, is based on the intensity of relationships any given person has with a museum. While hierarchical in its presentation, the pyramid depicts the fact that at any given time or in any given situation, the intensity of the relationship can't be all-encompassing. There are always those more engaged and those more passive and a good museum needs to have a variety of approaches to address all aspects of the ladder.

The idea behind Figure 1 is to show that while museums are considered to be part of any social and cultural space, anyone who is just generally aware of a museum could be considered its public. The concept 'public' referrers to a large, unattached set of people. Those people form the base set from which the museum, or the people themselves, are able to select. The awareness is needed in order to be able to move on to become more engaged. The more people are engaged with the museum, the higher up they are situated on the pyramid. However, while people's engagement with the museum increases, the number of those people decreases, as it is by no means feasible and also not necessary to try to engage everyone very deeply with the museum.

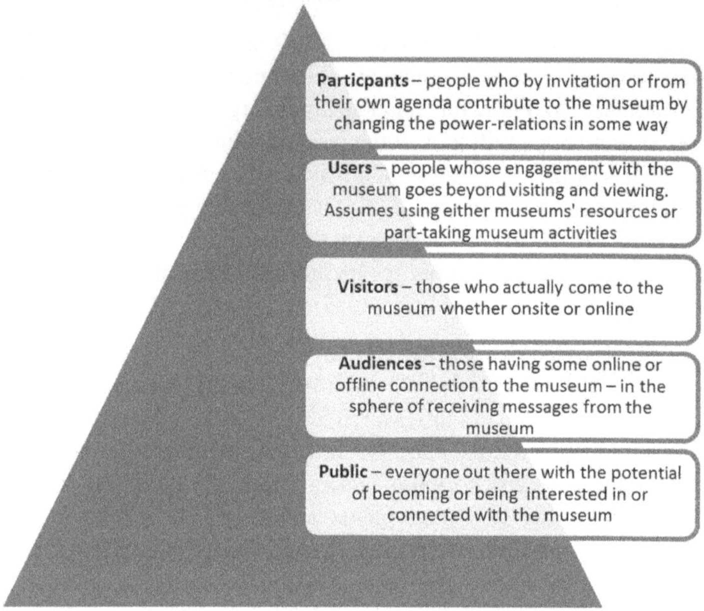

Figure 1: Progression of people in and around museums from public to participants

'Audiences' are groups of people who are already aware of the museum's messages. In today's networked world, museum audiences may actually never enter the doors of a museum. Museums can also spread ideas, messages and content via traditional or online media, but a museum's messages can also be spread by other people, for example in the form of viral messaging in online channels, or word of mouth by experienced museum goers. In contemporary society, where attracting visitors and increasing visibility of the museums are crucial for the sake of the existence of these institutions, 'new audiences' gain particular significance. This may mean that the museums start thinking about their audiences in a different way, seeing and acknowledging that not all members of the public are necessarily museum audiences and also that they do not form a unified group who can be reached with the same kind of messages and communication channels. Elsewhere throughout the book we have used Nico Carpentier's (2007, 2011) model of access, interaction and participation. The audiences have to have the minimal – access to the museum or its resources.

'Visitors' are the most traditional groups, discussed extensively in museum research. However, the concept of 'visitors' is in many ways problematic, as it traditionally denotes the people who enter the museum to pay it a visit. However, today, it is becoming harder and harder to distinguish when a visit starts or ends as a lot of prior engagement with the museum takes place in online spaces and often the visiting experience itself is carried across the museum doors to the digital realm, either as a single visit to the museum's web page or to form some kind of more permanent relationship with the museum. In some contexts, it would therefore make sense to develop the notion of 'users', as they would be easily recognised as people who use the online resources/spaces of a museum. In this book, however, 'users' are also people who use museum resources either on-site or online, using the museum for meaning-making (Runnel, Lepik, Lotina, forthcoming). This means that while internet studies would focusing on online users and would conceptualise usage as such in online space only, in this book we see also a museum site as a place for usage. The concept of 'users' matches well with the notion of interaction (Carpentier, 2007; 2011) as this assumes personalised and engaged activity with the museum with some limited feedback and dialogue opportunity.

'Participants', whom we could define as people with whom the museum is willing to share a small amount of decision-making power (as also in Carpentier, 2007; 2011) are probably the most desirable group with which a museum is looking to engage. However, 'participants' also demand most attention and continuous dedicated work to maintain a steady relationship. This means that they are also the smallest group in any given situation. In addition, while the museum might desire such a deep engagement in some situations, at other times museums might seriously limit participation and be happy with people as users.

When we look from the perspective of Morrone (2006), any kind of engagement in cultural activities could be considered cultural participation. This means that already at the level of the audience people participate in consuming cultural messages from a museum. While the ladder above (Figure 1) stresses that participants are people who have invested in museum activities and with whom a museum shares decision-making power, in the next section we are going look at the cultural participation framework and examine several conditions of possibility that may have an impact on whether the public is willing to become 'audiences', 'visitors' or even 'participants'. We need to acknowledge that these conditions of moving up or down the ladder are also a two-way street in terms of how it connects audiences and institutions. The conditions for participation can stem from the institutions (e.g. access, offering opportunities for interaction or fostering activities of participation), but the conditions of possibility may also originate from the visitors (e.g. various types of capital, including education and more specifically information literacy as well as the social identity of visitors influencing whether a person becomes more or less engaged with a museum). In the following, we will briefly discuss engagement from the aspect of both institutional and personal starting points – an outline originally provided by Krista Lepik in her doctoral thesis on *Governmentality and Cultural Participation in Estonian Public Knowledge Institutions* (Lepik, 2013), which we develop further in order to introduce or outline some aspects that are specifically relevant in the current discussion.

3. Conditions of participation stemming from museums

As mentioned above, the conditions of possibility for cultural participation can be derived from the access, interaction and participation (AIP) model (Carpentier, 2007; Goodnow, 2010; Carpentier, 2011). This model (see this volume, 89–106, Lotina) has inspired discussions in several chapters in this volume (see this volume, 35–53, Pruulmann-Vengerfeldt, Runnel; 91–108, Lotina; 131–148, Tatsi, 149–160, Tatsi, Aljas) and in this concluding article we will review how museums can support or hinder development from public to participants by acknowledging different elements in this model. It is very important, however, to understand that different modes of engagement as well as acknowledgement of different groups within a museum is very context specific. While a museum can be very open and invite participation and participants in some areas of its activities, it may very strictly limit access and hence also engagement with a museum in some other aspects. The conditions of participation discussed below are very context specific and can quickly change over time, over groups of people as well as over specific situations. Different exhibitions, education programmes or participatory initiatives can alter the museum approach vastly.

Access may seem at first glance to be a basic premise for cultural participation, but it has not always been accepted as a basic condition for participation, as is seen now. Enabling or disabling access or interaction in a public knowledge institution is largely a matter of choice following the tacit rules of the institution and thus made either willingly or not. Whether access is provided or not thus stems from organisational culture, which could be understood as "the glue that holds an organization together through a sharing of patterns of meaning" (Siehl and Martin, 1984: 227, cited by Carpentier, 2011: 218). According to this understanding, organisational culture "focuses on the values, beliefs, and expectations that members come to share" (ibid.). This means that when deciding which modes of cultural participation are accepted or not, museums are making ideological choices. These choices are carried in the organisational culture as acknowledged (or sometimes not acknowledged) ways of doing things.

Both access and interaction can be fostered or limited in several ways, ranging from enabling or limiting the physical access of disabled people to the museum; the inability, or possibility, to consume certain forms of culture because of, or despite, some sensory problem, to enable or block access to technological equipment needed in order to consume or produce culture (Weisen, 2013). Other ways of limiting access can be providing information in the dominant cultural language only and excluding minority languages or English, the *lingua franca* of tourists. A museum might ignore the fact that a considerable amount of its visitors might be shorter than 150 centimetres thus making a lot of exhibited content inaccessible for children. The role of information and communication technologies (ICT) has in such cases been emphasised (Ward, 2010), often in terms of the digital divide (e.g. van Dijk, Hacker, 2003), but also in terms of pointing to potential mental barriers related to ICT usage (ibid.). Acknowledging that people with different sensory abilities are part of the museum's desired audiences, makes it possible to address various issues of access and to question the quality of the access provided.

With particular importance in museums, provision of access to information with the help of ICT is also grounded in multiple layers of information content. Firstly it involves basic information about the institution (e.g. opening hours, facilities); secondly, metadata about various information resources in other museums or public knowledge institutions more generally can be provided; thirdly, a museum can provide access to an information system of its own; fourthly, in the case of sufficient resources and perceived need, a museum can provide access to its digitised content (Maier, 2002).

Of all these layers, the first three focus on the usage of the museum, the possibility to access information, for example whether the museum is open when one needs to go there, whether the required material is physically available in a museum, or whether it is possible to view material online regardless the physical access to the museum that preserves the physical object. The fourth layer is access to information provided online, allowing the constraints related to visiting a museum to be crossed. These four layers of access do not solve the problem of division between inclusion in, and exclusion from, information flows, but to some extent the exclusion is reduced within the facilities of museums either by providing access to materials on site or online, or by allowing people to use technological equipment provided by these institutions.

The development of ICTs is important in the museum context as it has increased the possibility of accessing the information. Yet, it has also forced museums to reconsider the visitor in the online space and think what is provided through online channels, how it is provided, and how to engage people to become regular users of online resources. Some of these constraints have also been discussed in Aljas and Pruulmann-Vengerfeldt (this volume, 163–183). Providing increasing amounts of information online highlights an important individual attribute, namely information literacy, which is becoming relevant in the context of cultural participation. However, although the individual's capacities become foregrounded, this does not exempt museums from the responsibility of making materials as accessible as possible through all means.

Provision of access with sophisticated technologies has corresponded well to the working logics of museums, even helping to solve some fundamental conflicts, for example digitisation allows the institution to provide access to images of rarities, while the original object can be preserved in the depository (this volume, 163–183). Interaction and technologies enabling interaction have complemented, but sometimes also challenged, the logics of access provision with the possibility to provide new content and discuss existing content (Carpentier, 2007). Web 2.0 has been seen to provide an "architecture of participation, a built-in ethic of cooperation" (O'Reilly, 2005: online), thus fostering the interaction. In museums, social media has been seen as a tool to increase interaction with visitors (Russo et al., 2010), and folksonomies or tagging projects in particular have been suggested to "foster and maintain links with specialized groups like volunteers and docents, or support the work of teachers and students" (Trant, Wyman, 2006: 3). While the possibility of outreach or instruction is generally seen as acceptable for museums, social tagging has been viewed with certain cautiousness. This cautious attitude is also familiar in Estonian museums, reflected by a study conducted by Pruulmann-Vengerfeldt and Aljas (2009; this volume 163–183), who conclude that many museum experts feel that as existing cataloguing systems and database structures have worked for museums for nearly a hundred years, they should continue to do so. The question of interaction is therefore related not only to technology, as it may appear at first glance. Through attitudes and values interaction is also related to organisational culture and from there to underlying ideologies and governance rationalities in museums.

The conditions for the narrowest concept, 'participation', are discussed in the methodology chapter in this book (this volume, 55–74, Pruulmann-Vengerfeldt, Runnel, Aljas). For Carpentier (2011) participation means the ability to co-create content, take part in evaluating and deciding processes on technological or process levels. What it means for the museums to be fostering participation is discussed in detail in all the other articles of this book, hence we will now proceed to take a look at the conditions supporting or hindering people in developing relationships with the museums.

In the next section, we will look at the personal attributes that could condition the possibility of participating in museum activities and support the change from general member of the public to museum participant. While there can be an infinite number of conditioning attributes, we focus on two sets of key factors: firstly, capitals, including education and information literacies, and secondly the ways in which identity is related to people's engagement with museums. Various aspects outlined in the following do not constitute a conclusive list that explains different modes and scope of engagement. Rather, these aspects form a set of significant elements that could be taken into consideration when trying to understand how well a museum has been able to relate to people and how well it has addressed diversity. This means that these elements are not conditional for becoming visitors/participants, but enable the mapping of different tracks along which a person might move when becoming more or less engaged with a museum.

4. Participation factors stemming from people

4.1. Capitals as personal attributes related to museum engagement and participation

When moving from the above-described ladder (Figure 1) in becoming increasingly engaged in museum and eventually becoming a participant, the individual attributes can play an important role in defining both the governing conditions, but also availability of the resources to spend. While our goal is not to give a defined list of attributes conditioning museum engagement, the different aspects outlined below help us to understand some of the different conditions influencing people at any given point of time or in any given situation.

One of the traditional starting points has been considering measurable variables such as income or standard of living. The most common assumption in this case is that "democracies will more likely exist in richer rather than poorer countries" (Krishna, 2008: 1); a considerable amount of studies confirming this assumption have been outlined by Krishna (2008). Sometimes, researchers have even gone as far as attempting to propose living standards above which democracy might survive (for example, "in countries with per capita incomes above $4,000" (Przeworski et al., 2000, cited by Krishna, 2008). As democracy has been seen as a "sort of luxury of good" (Barro, 1996: 24, cited by Krishna, 2008: 3), cultural participation of any kind – consumption of cultural goods or active creation of additional value in the cultural domain can be seen as relatively resource hungry. The level of income has also been related to a feeling of existential security, leading people "to shift their emphasis from survival values toward self-expression values and free choice" (Inglehart et al., 2008: 266), allowing emphasis of participation as well. And while being involved with a museum might not always require money, it requires a willingness to place emphasis on cultural participation.

With a certain caution we can treat the level of income as financial or economic capital, and thus consider other 'capitals' as well. Bourdieu offers in his *Distinction* ([1984] 2010) a view that education and cultural capital can be treated hand in hand with economic capital in order to study the cultural preferences of various social classes and professions. While it is not our aim to discuss whether there is a certain level of economic capital required for museum participation, we do believe that this can be an indicator of the resources necessary.

When discussing participation, social capital (also an existing concept in the works of Bourdieu, closely related to the notion of symbolic capital (Siisiäinen, 2000)) has been analysed within the framework of civic participation (Putnam, 2000). The impact of social capital on civic participation has been questioned by Hooghe (2003) who, after drawing on the work of Putnam (2000), compared the reasons for the decline of civic participation in the United States of America, and Belgium. Hooghe's work has shown that the impact of social capital can be highly dependent on context. Hooghe shows significant differences between these two countries as "with the notable exception of religion and secularization, none of the factors that are cited in the literature as responsible for the decline of participation levels are significantly related to the intensity of participation" (Hooghe, 2003: 55). What is particularly interesting is that Putnam (2000) too is quite careful about interpreting his results, as shown for example in marital and parental status (compared by Hooghe, 2003), as he eventually states that, "apart from youth- and church-related engagement, *none* of the major declines in social capital and civic engagement that we need to explain can be accounted for by the decline in the traditional family structure" (Putnam, 2000: 279). As the body of literature has shown, apart from the impact of social capital, there is a lack of a common understanding of the components of social capital. In this article, social capital is seen as being related to basic categories such as trust or trustworthiness (Offe, Fuchs, 2002: 190; Ostrom, Ahn, 2009: 20), and being part of associations or networks (Siisiäinen, 2000; Wuthnow, 2002: 63; Ostrom, Ahn, 2009: 20) – following therefore the approach of Putnam (2000) who also defines social capital as "connections among individuals – social networks and the norms of reciprocity and trustworthiness that arise from them" (Putnam, 2000: 19).

If we understand museum-going and becoming involved in museum activities as a potentially collective/social activity, social capital functions as a prerequisite. Inspired by Simon (2010) Tatsi (2013) describes how the social museum means that interaction is not only important between museum and person, but also that the museum becomes as social space to mediate networked activities between visitors.

In a potential situation in which many new ideas have been proposed, political capital may start to foster or hinder becoming engaged in museum activities. Just as is the case for social capital, political capital can also be defined in multiple ways. From the perspective of this article it is necessary to point to the proximity between concepts of social capital and political capital, as these "forms of capital accumulate in relational ties" (Nee, Opper, 2010: 2107). At the same time, political capital differs from social capital because it "has the additional feature of being linked to the positional power of the politician, and thus it is rooted in institutional structures of the political order" (ibid.). In the case of museums that would like to define themselves as politically neutral or ideology free, the political capital becomes debatable. Yet, as in the light of minimalist-maximalist versions of democracy it is possible to move beyond institutional politics, and define "the political as a dimension of the social" (Carpentier, 2011: 17), political capital can obtain a somewhat broader meaning in museums and libraries. For example, a respected member of the local community of some ethnic minority, actively representing the community in a museum, can draw on her or his political capital. Both the positional power and representative function in this case support participation in culture, and help to legitimise certain ideas, viewpoints or activities.

The idea about cultural capital, which according to Bourdieu is mainly defined through formal and informal education as a prerequisite for participation, is old and can be found in the works of Enlightenment philosophers (Rousseau, J. S. Mill) tightly related to the assumption of the educative function of participation. Pateman (1970), who has analysed the development of participatory theory, has on this point referred to Rousseau's notion of the "self-sustaining" (Pateman, 1970: 25) participatory system in which "the more the individual citizen participates the better able he is to do so" (ibid.). 'Learning democracy', starting from the local level and ending perhaps even at the level of national government, or conveying ideas about participation from one field to another (Pateman (1970) is also related to the concept of education, or in a way to 'literacy' in the sense of becoming capable of participating in democracy.

4.2. Information literacy as personal attribute related to museum engagement and participation

While information literacy is certainly part of any good education in the modern world, in this article, as a condition of the possibility stemming from visitors, information literacy could be treated as a form of capital of its own. In varying contexts, particular literacy-related concepts have also been seen as prerequisites for participation in civic society. In terms of media literacy, for example, the ability to create content, and the competency to actively participate in social processes have been seen as fostering participation in democratic processes (Runnel, 2009; Ugur, 2010). While initially 'information literacy' in the context of cultural institutions was related to knowledge of information resources, it has gradually moved beyond the work setting and started to serve a wider function (Bawden, 2001: 230). The extension of information literacy "to the functions of citizenship" – e.g. "beyond information literacy for greater work effectiveness and efficiency, information literacy is needed to guarantee the survival of democratic institutions" (Owens, 1976, cited by Bawden, 2001: 230) started in the 1970s. By the 1990s, as Sanna Talja and Annemaree Lloyd point out, the "idea of empowering individuals through teaching and adoption of information acquisition skills and competencies" (Talja, Lloyd, 2010: X) had gained the status of a "powerful way of thinking about information literacy" (ibid.).

To be able to participate in museums one needs to be sufficiently knowledgeable about the institution, participation possibilities, boundaries that might allow or disallow participation. Some of these possibilities are made explicit for visitors, for example in terms or rules of usage, or through the afore-mentioned modes of informing, yet some remain vague and may even be confusing for the staff if these possibilities are not formulated clearly enough. When we focus on cultural participation as the *consumption of culture* (for example reading a book, attending an exhibition, etc.) we can find quite clear instructions that directly or indirectly support the consumption of culture. However, focusing on cultural participation as the *production of culture* (for example, collaborating with the museum, contributing tangible or intangible cultural heritage) is both a novel and intriguing step, especially when this production is considered to be entrusted to visitors who are often seen as amateurs in culture.

4.3. Identity and engaging with museums

In addition to capitals, people's identity is a potentially explanatory aspect in understanding how and why people engage with museums. Identity as a concept can in itself be lent to a whole range of books: we can talk about social identity, ethnic, national, racial, or gender identity, cultural identity, personal or self-identity, collective identity, identity politics, etc. (e.g. Giddens, 1991; Jenkins, 2004; Alcoff, 2006). Hence the short discussion here does not attempt to cover all aspects, but just points to a few of those we consider noteworthy in order to understand the museum visitor.

In the museum context, identity work has been analysed from the perspective of national (e.g. Coombes, 2012), ethnic (e.g. Crooke, 2007) or racial identity (e.g. Bennett, 2005); however, very often the focus has been on how museums can help or hinder the individuals coming to the museum. In the context of this article, we will look at how identity would support building the relationship between an individual and the museum. When it comes to relationship building, in museum studies people's visitor identities have generally been discussed and seen in close relation to the visitor's motivation for visiting the museum (Falk, 2009; 2011; Pitman, Hirzy, 2010). Within the framework of our paper, the same identity discussion can and should be extended across all of the above-mentioned levels from public to participants. Museums can be seen as a more or less important part of these people's identity building as soon as they become aware of the museums as such. Even deliberately not going to a museum can be part of someone's identity construction. Understanding this allows us to discuss identities bound to diverse relationships between individuals and the museum, extending beyond the narrower notion of 'visitor identities'. Nevertheless, the 'visitor identities' can still be treated as a starting point for further theorisation.

Falk (2009; 2011) has in his studies largely drawn upon identity theories summarised by Cooper (1999) and distinguished various identities as big "I" identities, which are enduring and deep, remaining fairly constant across our lives (Falk, 2011: 6); and small "i" identities which he sees as situated identities that represent responses to the needs and realities of the specific moment and circumstance (ibid.: 7). In doing so, he rejects the dominant framework where visitor researchers have focussed on "permanent qualities of either the museum" (ibid.: 2), "or the visitor" (ibid.), without paying sufficient attention to the particular "relationship that occurs each time a person visits a museum" (ibid.: 4). Instead, he proposed an approach which attempts to position visitors in identity-related categories originating in particular museum visits (Falk 2009). While Falk brings social elements to identity construction in relation to the museum, it is important to see that we could talk about situational identity (Carter, 2011: 300) – the aspect of the identity complex "given by one's position in particular interactional contexts" (ibid.). The situation can be triggered by the museum, the exhibition, the other people with the visitor, by the time of the day and many other factors (including those outside the museum – news about the museum for the public). Falk's significant contribution to visitor studies is his emphasis upon people's everyday behaviours, attitudes, and needs and how they shape visitor identities: "The type of identity that does figure prominently into the myriad everyday decisions in our lives, including leisure, are what I have called 'little "i" identities' – identities that respond to the needs and realities of the specific moment and situation" (Falk, 2009: 73). This brings fluidity into understanding the visitor and unbundles the fixed concept of visitors being and remaining the same. Coming from a marketing perspective, Falk's conceptual take on identity helps to understand the visitor almost as "marketing personas" (Isidoro, 2013). This has helped museums to overcome the notion of visitors as a stable concept with a stable and fixed understanding of the museum.

Falk looks at visitor motivations, helps to describe visitors, but also articulate their positions in relation to the museum: 'explorer', 'facilitator', 'experience seeker', 'professional/hobbyist' and 'recharger' (Falk, 2009: 64); later, 'respectful pilgrims' and 'affinity seekers' (Falk, 2011: 10) were added. These categories can be understood as ideal types, as visit motivations combine some mix of all these reasons (ibid.). Falk (2009; 2011) views visitor identities "that respond to the needs and realities of the specific moment and situation" (Falk, 2009: 73), including a visit to a museum. It is therefore important to see that the same ideal types can shift even when walking through the exhibition, let alone from when you plan your visit online to the time you exit with interactive materials for later exploration. According to Falk's view, these shifting identities influence the continuous cycle between visit expectations, and visiting satisfaction and memories gained from the museum (Falk, 2009). As we can see, for Falk primary interest is limited by a museum visit and is rich within these boundaries and in the framework of behaviourist marketing research; however, this is not good enough in our exploration of various forms and modes of people's engagement with museums for various reasons including how this relates to the aspect of identity. Firstly, in Falk's visitor motivations framework, museum visits remain unquestionably the site of primary interest, and thus the model is limited by a primary focus on the institution rather than visitors' lives (see also Dawson, Jensen, 2011: 131 for a thorough critique of this aspect). Secondly, this limits, as do the majority of museum visitor studies, his interest in people who already visit the museum. When trying to understand the visitor among the other modes of relationships a person can have with a museum, which is the primary interest of this article, our scale (Figure 1) also involves people who might never visit the museum in physical space, but who develop some kind of engagement with the museum institution; they might relate to the meanings or content, offered by the museum or might themselves contribute to the museum. In addition, when exploring these engagements, the museum institution remains inevitably central to our interest, but the theoretical focus shifts to the communication between the person and the institution.

Despite Falk's interest in everyday life context, his identity-based model has also been criticised for using a reductionist approach, where dealing only with situational identities does not enable us to notice the richness of people's identities (and when and where these actualise in the process of meaning-making across different social and personal contexts, situations, social practices). This means that when he focuses on the "i" in identity, the more fixed aspects might be ignored. Therefore, it is important to note that the aspects of identity that are usually important outside the museum, like socio-demographics or being a member of other communities of practice (including particular heritage communities), as well as more private aspects of identity, influence the relationship with a museum. Hence the focus cannot only be on the identity formed in relationship with the museum, but is also on the other dimensions that are inevitably present when thinking about museums, viewing, using or participating in or around museums. Hence sometimes, people who could not even be considered publics – for example those so far from the museum that they are in no communication loop – can receive mediated messages from the museum (for example, a friend relating something) that trigger them to become active participants. In this book we have discussed the example of the "Give the Museum A Day from Your Life" project, which aimed among other things to raise awareness of the importance of collecting everyday phenomena for an ethnographic museum. We believe that this campaign might have triggered aspects of people's identity that encouraged them to participate even if their prior museum-related situational identity experience was completely missing.

5. Conclusion

This paper approached a variety of relationships between individuals and museum institutions by considering capitals and identities as conditions for cultural participation. Apart from the capacities of individuals, we looked at the role of the museum in the formation of these relationships as a provider of access to museum contents and as a facilitator of interaction with individuals. Supporting the framework of cultural participation, this approach is comparable to Steyart's (2002) multidimensional approach to civic participation related to the digital divide, whereas civic participation is facilitated by access to information as well as a set of skills ranging from instrumental and structural (covered largely by capitals in our approach) to strategic (capitals and identities). This layered approach is useful when the affordances and motivations of individuals to participate in culture through museums are discussed, especially in situations where the door is opened wide yet people seem hesitant about entering (the museum).

We considered the complexity of conditions to cultural participation and kept in mind a mutual support between the aforementioned conditions: access, interaction, capitals (including information literacy) and identities. Capitals providing the necessary financial, cultural, social and political means support the formation and expression of identities (which help to form a certain kind of relationship with a museum). For example, participatory activities which require a lot of knowledge from a participant might require sufficient financial capital as a prerequisite of participation, whereas participatory activities which require a participant's time and skill depend on the potential participant's social capital or information literacy.

Information literacy as a certain kind of personal capital is linked to identity: identity is nourished by the knowledge necessary for participation (the experience of having sufficient knowledge for doing something in the museum might serve as the trigger to become more engaged with the museum), and vice versa – becoming information literate is supported by the identity of potential visitor. When considering the relationship between the museum and the individual, it is important to notice that a person's social identity and information literacy may also be influenced by the modes of access and interaction provided by museums.

Various conditions for cultural participation that we approach in this article, stemming both from individuals and museums, have previously been partially outlined in Lepik's doctoral thesis (2013). However, the ideas from this thesis, dealing with the narrower framework of museum-visitor relationship, were elaborated further in this article: we considered a wide array of relationships between the museum institution and the public, from those who have just a potential interest in the museum up to those who, as participants, have developed close ties with a museum and might be passionate in what they are doing in relation to the museum. Another important difference from Lepik that can be accentuated hereby lies in the approach to the visitors. While Lepik (2013) focused on more or less materialised articulations and modes of governance of visitors, the current article has paid a great deal of attention to more or less distinguishable groups of people with whom the museum can communicate. There may be some cautiousness related to the application of ladder-based approaches to visitors (see also Pruulmann-Vengerfeldt, Runnel, 2011), yet the plurality of repertoires to engage people in museum activities remains notable also on the basis of this article. Moreover, it allows a rather clear-cut mapping of various people on the basis of the intensity of their relationship to the museum – especially despite the initial discursive confusion which was experienced by the members of our research group, probably familiar to any researcher or practitioner attending some interdisciplinary event with the focus on museums.

References

Alcoff, Linda Martín (2006). *Visible Identities: Race, Gender, and the Self.* Oxford, etc.: Oxford University Press.

Association of College and Research Libraries (ACRL). (2000). Information literacy competency standards for higher education. Chicago, IL. Available at: http://www.ala.org/acrl/standards/informationliteracycompetency (accessed November 8, 2013).

Bawden, David (2001). Information and Digital Literacies: A Review of Concepts. *Journal of Documentation*, 57 (2): 218–259.

Bennett, Tony (2005). Civic Laboratories. Museums, Cultural Objecthood and the Governance of the Social. *Cultural Studies*, 19 (5): 521–547.

Bourdieu, Pierre ([1984] 2010). *Distinction*. London and New York: Routledge.

Carpentier, Nico (2007). Participation, Access and Interaction: Changing Perspectives. In Virginia Nightingale, Tim Dwyer (eds.). *New Media Worlds: Challenges for Convergence*. Oxford [etc.]: Oxford University Press, pp. 214–230.

Carter, Angela (2011). Identity: Social Psychological Aspects. In George Ritzer, J. Michael Ryan (eds.) *The Concise Encyclopedia of Sociology*. Malden, MA; Oxford; Chichester: Blackwell Publishing, pp. 300–301.

Carpentier, Nico (2011). *Media and Participation: A Site of Ideological-Democratic Struggle*. Bristol, UK and Chicago, USA: Intellect.

Coombes, Annie E. (2012). Museums and the Formation of National and Cultural Identities. In Bettina Messias Carbonell (ed.). *Museum Studies: An Anthology of Contexts*. Malden, MA; Oxford; Chichester: Blackwell Publishing, pp. 260–272.

Cooper, Catherine R. (1999). Multiple Selves, Multiple Worlds: Cultural Perspectives on Individuality and Connectedness in Adolescent Development. In Ann S. Masten (ed.). *Cultural Processes in Child Development*. New Jersey: Lawrence Erlbaum. Available at: http://www.bridgingworlds.org/pdfs/paper7.pdf (accessed July 19, 2013).

Cooper, Catherine R.; Jill Denner (1998). Theories Linking Culture and psychology: Universal and Community-Specific Processes. *Annual Review of Psychology*, 49: 559–584.

Crooke, Elizabeth (2007). *Museums and Community*. Abingdon; New York: Routledge.

Dawson, Emily; Eric Jensen (2011). Towards A Contextual Turn in Visitor Studies: Evaluating Visitor Segmentation and Identity-Related Motivations. *Visitor Studies*, 14 (2): 127–140.

Falk, John H. (2009). *Identity and the Museum Visitor Experience*. Walnut Creek, California: Left Coast Press, Inc.

Falk, John H. (2011). Reconceptualizing the Museum Visitor Experience. Who Visits, Why and to What Affect? ICOFOM Annual Meeting 2011. Available at: http://network.icom.museum/icofom/meetings/icofom-annual-meeting-2011/ (accessed November 8, 2013).

Giddens, Anthony (1991). *Modernity and Self-Identity. Self and Society in the Late Modern Age*. Cambridge: Polity.

Goodnow, Katherine (2010). Introduction: Expanding the Concept of Participation. In Hanne-Lovise Skartveit, Katherine Goodnow (eds.). *Changes in Museum Practice: New Media, Refugees and Participation*. London: Museum of London and Berghahn Books, pp. XXV–XXXVIII.

Hooghe, M. (2003). Why Should We Be Bowling Alone? Results from a Belgian Survey on Civic Participation. *Voluntas: International Journal of Voluntary and Nonprofit Organizations*, 14 (1), 41–59.

Inglehart, Ronald; Roberto Foa, Christopher Peterson, Christian Welzel (2008). Development, Freedom, and Rising Happiness: A Global Perspective (1981–2007). *Perspectives on Psychological Science*, 3 (4): 264–285.

Isidoro, Andrew (2013). Marketing Personas for Non-Profits: Design and Implementation. *Guardian Professional*, July 29. Available at: http://www.theguardian.com/voluntary-sector-network/2013/jul/29/marketing-personas-nonprofits-fundraisers.

Jenkins, Richard (2004). *Social Identity*. London and New York: Routledge Taylor & Francis Group.

Krishna, Anirudh (2008). Introduction: Poor People and Democracy. In Anirudh Krishna (ed.). P*overty, Participation, and Democracy: A Global Perspective*. Cambridge: Cambridge University Press, pp. 1–27.

Lepik, Krista (2006). Tartu mäluinstitutsioonide ühisportaali kontseptsioon. [The Concept of Common Portal of Tartu Memory Institutions.] Magistritöö [Master Thesis]. Tallinn.

Lepik, Krista (2013). *Governmentality and Cultural Participation in Estonian Public Knowledge Institutions*. Doctoral Dissertation. Tartu: Tartu University Press.

Lepik, Krista; Nico Carpentier (2013). Articulating the Visitor in Public Knowledge Institutions. *Critical Discourse Studies*, 10 (2): 136–153.

Lipponen, Lasse (2010). Information Literacy as Situated and Distributed Activity. In Annemaree Lloyd, Sanna Talja (eds.). *Practicing Information Literacy. Bringing Theories of Learning, Practice and Information Literacy Together*. Wagga Wagga, N.S.W.: Centre for Information Studies, Charles Sturt University, pp. 51–64.

Macdonald, Sharon (2006). Expanding Museum Studies. In Sharon Macdonald (ed.). *A Companion to Museum Studies*. Oxford: Blackwell Publishing, pp. 1–12.

Mason, Rhiannon (2007). *Museums, Nations, Identities: Wales and its National Museums*. Cardiff: University of Wales Press.

McLean, Fiona Combe (1994). Marketing in Museums: a Contextual Analysis. In Kevin Moore (ed.). *Museum Management*. New York: Routledge, pp. 232–248.

Morrone, Adolfo (2006). Guidelines for Measuring Cultural Participation. UNESCO Institute for Statistics. Available at: http://www.uis.unesco.org/template/pdf/cscl/framework/CUL_particip.pdf (accessed July 19, 2013).

National Museums Making Histories in a Diverse Europe. EuNaMus report No. 7. 2012. Linköping: Linköping University Electronic Press. Available at: http://liu.diva-portal.org/smash/get/diva2:573632/FULLTEXT01.pdf (accessed July 19, 2013).

Nee, Victor; Sonja Opper (2010). Political Capital in a Market Economy. *Social Forces* (University of North Carolina Press), 88 (5): 2105–2132.

Offe, Claus; Susanne Fuchs (2002). A Decline of Social Capital? The German Case. In Robert D. Putnam (ed.). *Democracies in Flux: The Evolution of Social Capital in Contemporary Society*. Oxford et al.: Oxford University Press, pp. 189–243.

O'Reilly, Tim (2005). What Is Web 2.0: Design Patterns and Business Models for the Next Generation of Software. Available at: http://oreilly.com/ web2/archive/what-is-web-20.html (accessed July 19, 2013).

Ostrom, Elinor; T. K. Ahn (2009). The Meaning of Social Capital and its Link to Collective Action. In Gert Tinggaard Svendsen, Gunnar Lind Haase Svendsen (eds.). *Handbook of Social Capital: The Troika of Sociology, Political Science and Economics*. Cheltenham, UK; Northampton, MA: Edward Elgar, pp. 17–35.

Pateman, Carole (1970). *Participation and Democratic Theory*. Cambridge: Cambridge University Press.

Pitman, Bonnie; Ellen Hirzy (2010). *Ignite the Power of Art: Advancing Visitor Engagement in Museums*. New Haven & London: Dallas Museum of Art & Yale University Press.

Pruulmann-Vengerfeldt, Pille; Agnes Aljas (2009). Digital Cultural Heritage – Challenging Museums, Archives and Users. *Journal of Ethnology and Folkloristics*, 3 (1): 109–127.

Pruulmann-Vengerfeldt, Pille; Pille Runnel (2011). When the Museum Becomes the Message for Participating Audiences. *Communication Management Quarterly: Časopis za upravljanje komuniciranjem*, 21: 159–180.

Putnam, Robert D. (2000). *Bowling Alone: The Collapse and Revival of American Community.* New York; London; Toronto; Sydney: Simon & Schuster Paperbacks.

Runnel, Pille (2009). *The Transformation of the Internet Usage Practices in Estonia*. Doctoral Dissertation. Tartu: Tartu University Press.

Runnel, Pille; Taavi Tatsi, Pille Pruulmann-Vengerfeldt (2010). Who Authors the Nation? The Debate Surround the Building of the New Estonian National Museum. In Simon Knell, Peter Aronsson, Arne Amundsen (eds.). *National Museums. New Studies from Around the World*. Routledge, pp. 325–338.

Russo, Angelina; Jerry Watkins, Lynda Kelly, Sebastian Chan (2010). Participatory Communication with Social Media. *Curator*, 51 (1): 21–31.

Scorrano, Armanda 2012. Constructing National Identity: National Representations at the Museum of Sydney. *Journal of Australian Studies*, 36: 345–362.

Scott, W. Richard (2001). *Institutions and Organizations*. London; New Delhi: Thousand Oaks; Sage Publications.

Scott, Monique (2007). *Rethinking Evolution in the Museum: Envisioning African Origins*. Abingdon; New York: Routledge.

Sfard, Anna (1998). On Two Metaphors for Learning and the Dangers of Choosing just One. *Educational Reseracher*, 27 (2): 4–13.

Siisiäinen, Martti (2000). Two Concepts of Social Capital: Bourdieu vs. Putnam. Paper presented at ISTR Fourth International Conference "The Third Sector: For What and for Whom?" Trinity College, Dublin, Ireland, July 5–8, 2000. Available at: http://dlc.dlib.indiana.edu/dlc/bitstream/handle/10535/7661/siisiainen.pdf (accessed November 8, 2013).

Simon, Nina (2010). *The Participatory Museum*. Santa Cruz, CA: Museum 2.0.

Steyaert, Jan (2002). Inequality and the Digital Divide: Myths and Realities. In Steven Hick, John G. McNutt (eds.). *Advocacy, Activism and the Internet*. Chicago: Lyceum Press, pp. 199–211.

Sztompka, Piotr (1999). *Trust: A Sociological Theory*. Cambridge: Cambridge University Press.

Sztompka, Piotr (2003). Trust: a Cultural Resource. In Grazyna Skapska, Annamaria Orla-Bukowska, Krzysztof Kowalsky (eds.). *The Moral Fabric in Contemporary Societies*. Leiden: Brill, pp. 47–66.

Talja, Sanna; Annemaree Lloyd (2010). Integrating Theories of Learning, Literacies and Information Practices. In Annemaree Lloyd, Sanna Talja (eds.). *Practicing Information Literacy. Bringing Theories of Learning, Practice and Information Literacy together*. Wagga Wagga, N.S.W.: Centre for Information Studies, Charles Sturt University, pp. IX–XX.

Tatsi, Taavi (2013). *Transformations of Museum-embedded Cultural Practice*. Doctoral Dissertation. Tartu: Tartu University Press.

Trant, Jennifer; Bruce Wyman (2006). Investigating Social Tagging and Folksonomy in Art Museums with steve.museum. A paper for the Tagging Workshop, World Wide Web 2006. Edinburgh, Scotland, May 22, 2006. Available at: http://www.archimuse.com/research/www2006-tagging-steve.pdf (accessed November 8, 2013).

Ugur, Kadri (2010). *Implementation of the Concept of Media Education in the Estonian Formal Education System*. Doctoral Dissertation. Tartu: Tartu University Press.

van Dijk, Jan; Kenneth Hacker (2003). The Digital Divide as a Complex and Dynamic Phenomenon. *The Information Society: An International Journal*, 19 (4): 315–326.

Ward, Janelle (2010). Purchasing or Protesting? Expanding the Notion of the (Online) Citizen Consumer. In Tobias Olsson, Peter Dahlgren (eds.) *Young People, ICSs and Democracy: Theories, Policies, Identities, and Websites*. Göteborg: Nordicom, pp. 35–50.

Weisen, Marcus (2013). Accessible Digital Culture for Disabled People. In Pille Runnel, Pille Pruulmann-Vengerfeldt, Piret Viires, Marin Laak (eds.). *The Digital Turn: User's Practices and Cultural Transformations*. Frankfurt am Main: Peter Lang Verlag, pp. 182–197.

Wuthnow, Robert (2002). The United States: Bridging the Privileged and the Marginalized? In Robert D. Putnam (ed.). *Democracies in Flux: The Evolution of Social Capital in Contemporary Society*. Oxford et al.: Oxford University Press, pp. 59–102.

Acknowledgements

The authors of this book are grateful to the different funding agencies mentioned across this book that have supported preparation, writing and publishing of the articles in this collection. Some of the articles here have been previously published elsewhere. We are grateful for the publishers for granting permission to reprint them in this collection. Authors are also grateful for the Estonian National Museum, Institute of Social Studies (formerly Institute of Journalism, Communication and Information Studies) of University of Tartu for their institutional support for the cooperation and research conducted in the framework of this project.

Who Authors the Nation? The Debate Surrounding the Building of the New Estonian National Museum, Pille Runnel, Taavi Tatsi, Pille Pruulmann-Vengerfeldt

Runnel, Pille; Taavi Tatsi; Pille Pruulmann-Vengerfeldt (2010). Who Authors the Nation? The Debate Surrounding the Building of the New Estonian National Museum. In Knell, Simon J.; Aronsson, Peter; Bugge Amundsen, Arne; Barnes, Amy Jane; Burch, Stuart; Carter, Jen (eds). *National Museums. New Studies from around the World.* London and New York: Routledge, pp. 325–338.

When the Museum Becomes the Message for Participating Audiences, Pille Pruulmann-Vengerfeldt, Pille Runnel

Pruulmann-Vengerfeldt, Pille; Pille Runnel (2011). When the Museum Becomes the Message for Participating Audiences. CM – *Communication Management Quarterly,* 6 (21), 159–180.

Handicraft Hobbyists in an Ethnographic Museum – Negotiating Expertise and Participation, Krista Lepik, Pille Pruulmann-Vengerfeldt

Lepik, Krista; Pille Pruulmann-Vengerfeldt (2013). Handicraft Hobbyists in an Ethnographic Museum: Negotiating Expertise and Participation. In Tomanic Trivundža, Ilija; Carpentier, Nico; Nieminen, Hannu; Pruulmann-Vengerfeldt, Pille; Kilborn, Richard; Sundin, Ebba; Olsson, Tobias (eds.) *Past, Future and Change: Contemporary Analysis of Evolving Media Scapes.* Ljubljana: University of Lubljana Press, pp. 267–280.

Facing the Death of the Author. Cultural professional's Identity Work and the Fantasies of Control, Nico Carpentier

Carpentier, N. (2011). Facing the Death of the Author. Cultural Professional's Identity Work and the Fantasies of Control. *MATRIZes*, 4 (2), pp. 183–204.

Identity Struggles of Museum Professionals: Autonomous Expertise and Audience Participation in Exhibition Production, Taavi Tatsi

Tatsi, Taavi (2011). Identity Struggles of Museum Professionals: Autonomous Expertise and Audience Participation in Exhibition Production. *Journal of Ethnology and Folkloristics*, 5 (2), pp. 65–80.

Democratising Collections through Audience Participation: Opportunities and Obstacles, Taavi Tatsi, Agnes Aljas

Tatsi, Taavi; Agnes Aljas (2012). Democratising Collections through Audience Participation. *The International Journal of the Inclusive Museum*, 4 (4), pp. 31–40.

Digital Cultural Heritage – Challenging Museums, Archives and Users, Pille Pruulmann-Vengerfeldt, Agnes Aljas

Pruulmann-Vengerfeldt, Pille; Agnes Aljas (2009). Digital Cultural Heritage – Challenging Museums, Archives and Users. *Journal of Ethnology and Folkloristics*, 3 (1), pp. 109–127.

Happily Lost in the Virtual Space? Access, Interaction and Participation in an Antwerpen Virtual Arts Museum, Nico Carpentier

Carpentier, Nico (2005). Vrolijk virtueel verloren lopen in het Muhka? [Happily virtually lost in the Muhka?] *Freespace Nieuwzuid*, 5 (17), pp. 86–103. Previous version was published in Dutch.

Increasing the Usability of the Museum: Four Studies, Pille Pruulmann-Vengerfeldt, Pille Runnel

Pruulmann-Vengerfeldt, Pille; Pille Runnel (2011). Increasing the Usability of the Museum: Four Studies. In: *Proceedings of the International Conference "Re-thinking Technology in Museums 2011: Emerging Experiences: Re-thinking Technology in Museums: Emerging experiences, 26–27 May, 2011 University of Limerick.* (eds.) Ciolfi, L., Scott, K., and Barbieri, S.: University of Limerick, pp. 243–252.

www.ingramcontent.com/pod-product-compliance
Ingram Content Group UK Ltd.
Pitfield, Milton Keynes, MK11 3LW, UK
UKHW021842100926
8556IPUK00019B/1006

9 783631 649169